MEMBRANE PHYSIOLOGY AND CELL EXCITATION

CROOM HELM BIOLOGY IN MEDICINE SERIES

STEROID HORMONES
D.B. Gower

NEUROTRANSMITTERS AND DRUGS
Zygmunt L. Kruk and Christopher J. Pycock

DEVELOPMENT, GROWTH AND AGEING
Edited by Nicholas Carter

INBORN ERRORS OF METABOLISM
Edited by Roland Ellis

Membrane Physiology and Cell Excitation

BRUCE HENDRY, M.A., B.M., B.Ch.
House Surgeon,
Addenbrooke's Hospital,
Cambridge
Former tutor in Physiology,
Merton College,
Oxford

A CROOM HELM BOOK
Distributed by
YEAR BOOK MEDICAL PUBLISHERS, INC.
35 East Wacker Drive, Chicago

© 1981 Bruce Hendry

This book is copyrighted in England and may not be reproduced by any means in whole or in part. Application with regard to reproduction should be directed to Croom Helm Ltd Publishers.

Distributed in Continental North, South and Central America, Hawaii, Puerto Rico and The Philippines by
Year Book Medical Publishers, Inc.
(ISBN 0-8151-4267-6)

by Arrangement with Croom Helm Ltd Publishers

Printed in Great Britain by
Biddles Ltd, Guildford, Surrey

CONTENTS

Preface

Part One: The Concept of Excitability   9
1. Introduction   11
2. Membrane Structure and Properties   14
3. The Basis of Excitability   26

Part Two: The Communication of Information   39
4. The Nervous Impulse   41
5. The Synapse   57

Part Three: The Acquisition of Information   71
6. Vision   73
7. Mechanoreception   87

Part Four: Information into Action   97
8. Skeletal Muscle Activation   99
9. Cardiac Muscle   112
10. Smooth Muscle   126

Part Five: The Clinical Importance of Excitability   137
11. Anaesthesia   139
12. Membrane Excitability and Disease   148

Index   157

Part One

THE CONCEPT OF EXCITABILITY

PREFACE

This book is intended for undergraduates studying the biological and medical sciences. The field of excitable cell physiology is one which is found quite baffling by a significant minority of these students. My aim here is to provide a brief introductory account, based on a conceptual approach, rather than on a mathematical or historical description. Once the student has grasped certain basic ideas concerning excitable cell function, the individual examples which follow fit into a well-defined pattern. No attempt has been made to give credit in appropriate measure to the many scientists who have contributed to this field. The further reading cited has been chosen with the reader alone in mind.

I would like to thank Tim Cripps for help and advice, also Charlie Tomson and Michael Hart for their careful reading of the manuscript. Finally I am indebted to Jenny Kenyon for her excellent typing of the work.

1 INTRODUCTION

1.1 The Excitable Tissues

Living organisms are able to respond to changes in their environment. Large multicellular animals, such as man, achieve this through the activity of a sophisticated arrangement of specialised tissues which allow for rapid and integrated responses to external stimuli. These are the excitable tissues. They include (1) sense organs, (2) peripheral nerves, (3) central nerves and (4) muscles. The sense organs acquire information about the environment; the peripheral nerves communicate it to the central nervous system where it is integrated with data from other sources, and the muscles initiate action based on the processed data.

This book is an introduction to the physiology of the excitable tissues. Its main theme is that the property of excitability results from a specialisation of the excitable cell membrane. Furthermore the type of membrane specialisation which gives rise to excitability is common to many excitable tissues and has been adapted to perform many functions.

1.2 The Cell Membrane

The functional unit of all living organisims is the cell. Each such unit is bounded by a cell membrane which is so thin as to be invisible under the light microscope, but electron microscopy suggests that it is about 5 nm thick (1nm = 1 x 10^{-9} metres). This membrane has many important properties, but for present purposes two are of particular interest. First it is selectively permeable; it allows some molecules to pass across it (into or out of the cell) more easily than others. Secondly it is capable of pumping substances into the cell from the outside medium. By these mechanisms the membrane provides a controlled intracellular environment, suited to the chemical processes of cellular function. For example, a cell may accumulate glucose despite a low extracellular concentration, and tends to maintain a constant internal glucose concentration.

In a complex organism each separate organ is composed of cells which are specialised in some way to perform a particular function. This specialisation involves the cell membrane. For example, the cells which line the gut are able to transport nutrients into the circulation because their membranes are specifically equipped with pumps to do this. Cellular excitability is almost entirely based on the properties of the cell membrane; it is the result of a particular type of specialisation. All tissues involved in the rapid acquisition, transmission and use of information possess this membrane specialisation.

1.3 The Excitable Membrane

Whilst all cell membranes are selectively permeable, and allow some molecules to pass across them more easily than others, an excitable membrane is, in addition, able to undergo rapid *changes* in its permeability to certain charged molecules or ions. These changes only occur as a response to specific stimuli. Thus in a pressure receptor, for example, membrane ionic permeability alters as a response to mechanical deformation. An excitable membrane is one which can be triggered by a specific stimulus to undergo rapid reversible changes in its ionic permeability. These changes result in a flow of ions across the membrane. The precise nature of the triggering stimulus, and the identity of the ions concerned, differ from excitable cell to excitable cell. Nevertheless a general phenomenon can be discerned in tissues as diverse in function as cardiac muscle and the retina. The ion flows across the excitable cell membrane are a mechanism of signal amplification. Many thousands of ions may pass into the cell due to a single stimulating molecule or event. One effect of these ion flows is to change the electrical properties of the cell. Because of this, the techniques of electrophysiology have provided most of the data on which our present understanding of excitation is based. Recently, other experimental approaches have also proved valuable. These include protein biochemistry, the manufacture of artificial membranes, spectroscopy and X-ray diffraction studies.

In the next two chapters the excitable membrane concept is explored in detail starting in Chapter 2 with the components of the membrane and its structure. The mechanism by which the membrane components confer upon it the property of excitability is discussed in Chapter 3. There follows a selection of examples of excitable tissues which illustrate the general concept of membrane excitability, and the versatility with which a common mechanism is adapted for a variety of functions. The nervous impulse and the synapse are considered first as these are the best understood examples of excitability. They are concerned with the rapid communication and integration of information within the organism. Next the acquisition of information by 'tuning' excitable membranes to specific types of stimuli is described with reference to the retina and mechanoreceptor organs. Then the way excitable membranes are used to produce action is considered in skeletal, cardiac and smooth muscle. Finally some clinical implications are discussed, with reference to anaesthesia, a procedure where cellular excitability is reversibly depressed, and to certain diseases involving a derangement of membrane excitability.

2 MEMBRANE STRUCTURE AND PROPERTIES

2.1 Membrane Composition

The major constituents of cell membranes are lipids and proteins. Lipid molecules consist of a polar headgroup to which is attached one or two long hydrocarbon chains or tails. In Figure 2.1 is the chemical structure of one such lipid, phosphatidylcholine, and below this a schematic diagram of the molecule which illustrates its important properties. The hydrocarbon tails of the molecule are *hydrophobic* — they tend to avoid coming into contact with water molecules. The polar headgroup on the other hand is hydrophilic and prefers an aqueous environment. If lipids are mixed with water then a number of structures are formed with the common feature that the hydrocarbon tails congregate together to protect themselves from contact with the aqueous medium. In Figure 2.1 two of these are illustrated — a spherical micelle and a monolayer on the water surface.

Figure 2.1: The Chemical Structure of a Typical Membrane Lipid

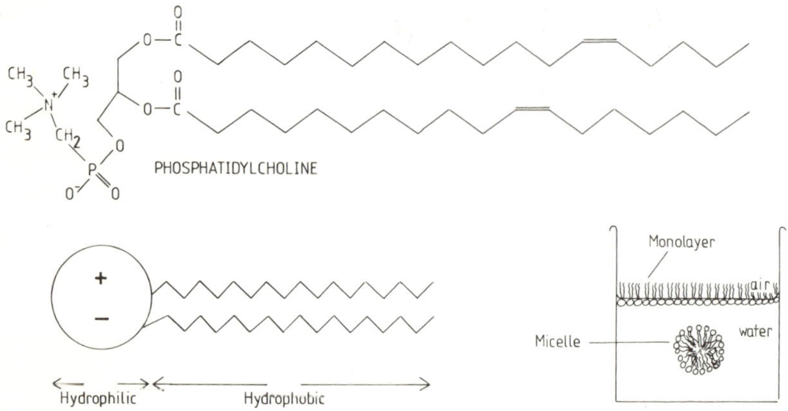

Proteins are formed by linking amino acids in long unbranching chains which are folded into an immense variety of shapes and sizes. The twenty or so amino acids are characterised by their different side chains which may be hydrophilic or hydrophobic. As the protein folds

it can hide its hydrophobic amino acids in the interior, and this type of folded structure is clearly suitable for proteins which operate in an aqueous environment. However *membrane* proteins may fold so that they have both hydrophobic and hydrophilic areas of surface like the membrane lipids. The lipids and proteins interact in the membrane so as to segregate hydrophobic and hydrophilic areas of surface as far as possible and this is the major factor determining membrane structure.

2.2 The Lipid Bilayer

Figure 2.2: The Lipid Bilayer

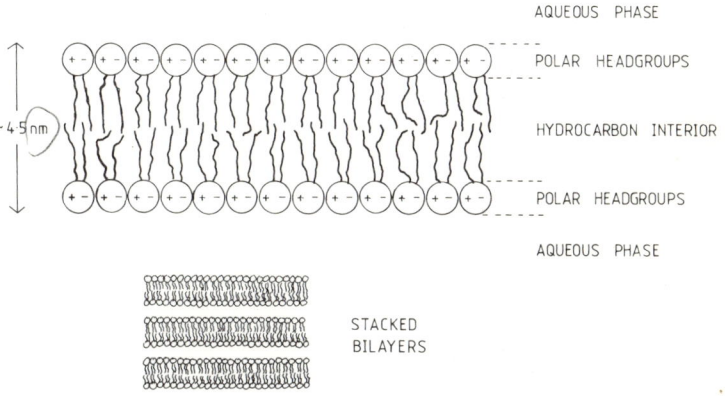

In those cell membranes which have been studied in detail the lipid molecules are found to be arranged in a bimolecular leaflet or bilayer as illustrated in Figure 2.2. Note that an unbroken surface of this nature successfully protects the hydrophobic lipid tails from the aqueous environment, and the bilayer is therefore likely to be a stable structure. The strongest evidence for the bilayer model is obtained by X-ray diffraction studies of stacked biological membranes. X-rays are of sufficiently short wavelengths to resolve the detailed membrane structure in a way that light rays cannot. Even so the technique is only accurate if a preparation of stacked membranes (as shown at the bottom of Figure 2.2) is used, so that the signal from each membrane can be added. The most appropriate preparations are the myelin membranes wrapped around the nerve axon and the layered disc membranes in the rod cells of the retina. The X-ray studies give the profile of *electron density*

across the membrane as shown in Figure 2.3. Below the profile is a diagram of the bilayer arrangement of lipid molecules and the two are consistent as the hydrocarbon lipid chains are of low electron density compared to the polar headgroups. The methyl groups at the end of the hydrocarbon tails have the lowest electron density of all. They appear to reside in the middle third of the bilayer giving rise to the central well in the electron density profile.

Figure 2.3: The Profile of Electron Density Obtained by X-ray Studies on Myelin Membranes. Note the agreement with the bilayer model shown below.

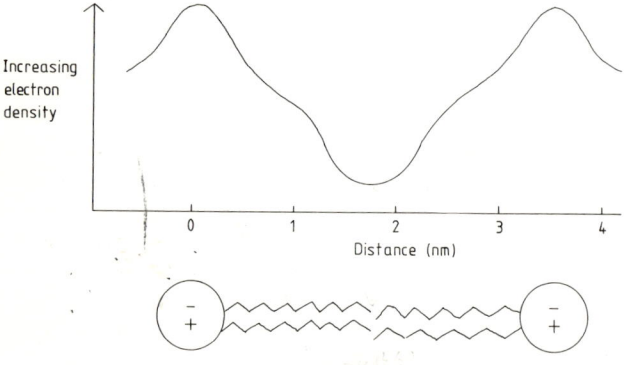

It is now possible to make artificial lipid bilayers which are amenable to close quantitative study and have confirmed the bilayer model of the cell membrane. For example electron micrographs of stained lipid bilayers closely resemble similarly-treated cell membranes. A measurement of considerable interest is the electrical capacitance of the bilayer as this can be used to give an estimate of bilayer thickness for comparison with the cell membrane. The hydrocarbon interior of the lipid bilayer is a continuous sheet of low dielectric constant (ϵ_h) which is expected to behave as a parallel plate capacitor whose capacitance (C) should be given by the relationship

$$C = \frac{\epsilon_h A}{d} \qquad (2.1)$$

where A is the area of bilayer and d is its hydrocarbon region thickness. The capacitance of artificial bilayers is found to lie close to 0.4 μF/cm^2 from which their hydrocarbon thickness d is estimated at 3 nm (30 Å), in agreement with X-ray studies of both artificial bilayers and cell membranes. Capacitance studies on cell membranes are complicated by the fact that the membrane capacitance is dependent on the frequency of the signal used to make the measurement. The high frequency capacitance of the squid nerve membrane is close to 0.4 μF/cm^2 and this measurement confirms the bilayer model. At low frequencies, however, cell membrane capacitance rises to about 1 μF/cm^2 and it appears that the bilayer structure must be interrupted by regions of higher dielectric constant which contribute to the capacitance at low frequencies but are silent at high frequencies. It seems likely that these areas are proteins in the membrane.

The excitable cell membrane functions largely by virtue of its selective permeability to ions, and numerous studies of the permeability properties of the pure lipid bilayer have been performed. The bilayer is highly impermeable to small ions (orders of magnitude less permeable than any known cell membrane) as might be predicted from its structure. To cross the bilayer an ion must enter the hydrocarbon interior at one interface and leave it at the other. This is a most unlikely event as (1) the transfer of a water-soluble ion into a hydrocarbon environment is energetically unfavourable and (2) it is highly unfavourable to place an electric charge in a low dielectric medium close to a conducting high dielectric medium — the aqueous phase. The latter is known as the *image potential* barrier. The bilayer is therefore an effective block to ion flow, and the permeability of excitable membranes must be due to other structures which somehow provide a pathway for ions to cross — again these are thought to be membrane proteins. The bilayer permeability to water is much higher than that to ions as the image potential barrier is not encountered by a neutral molecule. Water flux across the bilayer is inhibited by the unfavourable transfer of a water molecule into the membrane interior, but the bilayer water permeability is close to that found in most cell membranes. Water is almost always in equilibrium between intercellular and extracellular compartments. The molecules which cross a bilayer most easily are small and hydrophobic as they rapidly enter the hydrocarbon bilayer interior. Ethanol is an example of this group and, as expected, cell membranes are highly permeable to these substances.

It is important to stress that the lipid bilayer is not a static structure and allows its components considerable mobility. Each lipid mole-

18 *Membrane Structure and Properties*

cule can diffuse freely in the plane of the bilayer but cannot readily 'flip' from one interface to the other. The proteins which reside in the membrane also have considerable two-dimensional mobility. For example there is a phenomenon known as 'lymphocyte capping' in which a population of membrane proteins change from a random distribution throughout the membrane area to congregate in one small zone. Such changes in lateral mobility may be important in physiological signalling. In a sense the lipid bilayer behaves as a rather viscous two-dimensional liquid.

On first inspection the contribution made by the lipid bilayer to membrane excitability appears small because it exhibits none of the ion permeabilities found in excitable membranes. In fact the lipid bilayer is vital as it provides the appropriate environment in which membrane proteins can reside and function. The proteins endow the membrane with its specific ionic permeabilities, but to do this they must be 'sealed in' by the impermeable bilayer matrix. If the bilayer was freely permeable to ions then every ionic species would rapidly come to equilibrium between intracellular and extracellular compartments. It is vital to excitability that this does not occur.

2.3 Membrane Proteins

Figure 2.4: Membrane-protein Interactions. Protein A is an extrinsic membrane protein while proteins B and C are intrinsic.

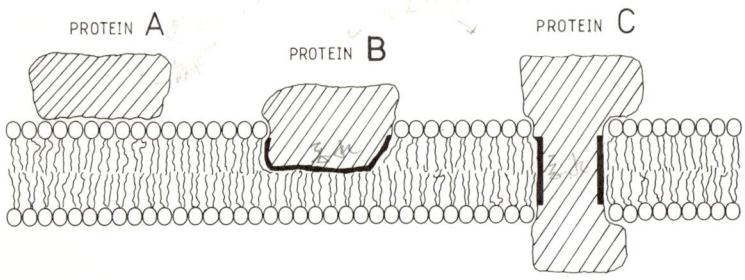

Proteins typically make up about one-third of the dry weight of a cell membrane, but this fraction is extremely variable. The properties of a protein are determined by the sequence of amino acids in its backbone. As there are over 20 different amino acids and a protein may have over 100 amino acid residues in each chain, the number and variety of possible structures is immense. Thus proteins can possess highly selective properties, and one group of proteins fold into conformations which are most suited to a membrane environment; these are the membrane proteins. Some of these proteins can be extracted from membrane fragments by gentle shaking in ionic solutions whilst others can only be isolated from the membrane by vigorous treatment with detergents. The former are known as *extrinsic* membrane proteins, the latter as *intrinsic* membrane proteins. Some possible arrangements of proteins in the membrane structure are shown in Figure 2.4. The three proteins (A, B and C) illustrate three possible examples of bilayer-protein interaction. Protein A makes contact with the lipid headgroups and surrounding aqueous environment, and has a completely hydrophilic surface to suit its surroundings. It can easily be extracted from the membrane and is an example of an extrinsic membrane protein. Protein B is partially inserted into the hydrocarbon bilayer interior and so has an environment which is partly hydrophilic and partly hydrophobic. Its surface is accordingly partly hydrophilic and partly hydrophobic (as indicated by the shading). Its position in the membrane has been determined by these surface properties in that only the hydrophobic surface is inserted into the bilayer interior. Protein C completely spans the bilayer and accordingly has a cylindrical area of surface which is hydrophobic (as indicated), and hydrophilic regions at each end of this. Proteins B and C are difficult to extract from the membrane as their hydrophobic surfaces would thereby be exposed to water; they are therefore examples of intrinsic membrane proteins.

Proteins of type C which actually span the bilayer are thought to be of great importance in excitable membranes as they are clearly well suited to providing a means by which ions may cross the membrane without passing through the bilayer. Two examples of evidence for protein molecules which are in contact with both sides of the membrane may be mentioned here. The first is in the red-blood-cell membrane and the second is in the 'purple' membrane of the bacterium *Halobium halobius*. Bretscher has used a radioactively-labelled molecule known as FMMP which binds to amino groups on hydrophilic protein surfaces, in order to study the proteins of the red-blood-cell membrane. If FMMP is incubated with a suspension of red blood cells then a number of membrane proteins become radioactively labelled. The proteins are then extracted and separated and an estimation made of which locations on the amino acid sequence of a given chain are radioactive, as illustrated in Figure 2.5. As FMMP does not usually penetrate into the cell (experiment A), this result indicates that these portions of the chain were exposed on the external surface of the membrane. The second part of the experiment, illustrated in part B of Figure 2.5, involves making the cell membrane leaky to FMMP so that it can label membrane proteins from the inside also. On extraction two of the proteins originally labelled by external application are found now to be labelled *at extra sites* on the same chain. These proteins are therefore concluded to have spanned the membrane as illustrated. The larger of these two proteins is known as component A and may be involved in providing the red-cell membrane with a high chloride ion permeability. The second example of direct evidence for a spanning structure is the pigment protein bacteriorhodopsin of *Halobius halobium* which is present in such large concentrations as to give the membrane a purple colour. Because of the high concentration of this single protein in the membrane a picture of its structure has been obtained by electron beam scattering. This demonstrates that each molecule possesses seven cylindrical units which completely span the membrane, and these are thought to be involved in the transport of protons (H^+) across the membrane.

Figure 2.5: Evidence for a Protein Spanning the Red Blood Cell (RBC) Membrane. In experiment A, FMMP does not enter the cell. In experiment B the cell membrane has been made leaky so that FMMP can enter.

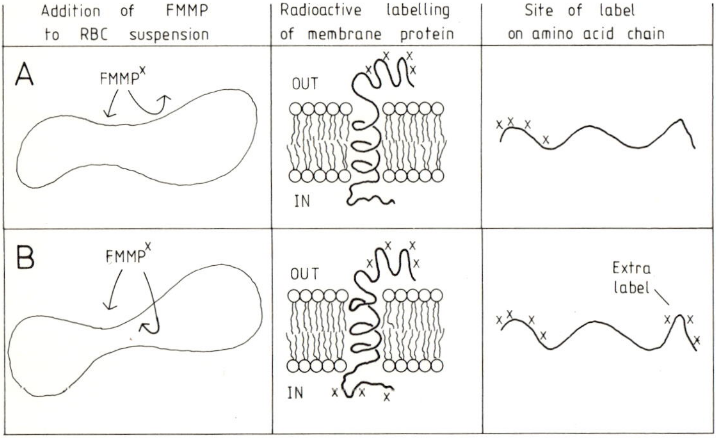

Of the large amounts of protein in an excitable membrane only a small proportion appears to be directly involved in ion flow. These true 'excitable proteins' are too scarce to be investigated by the techniques mentioned above. Therefore their characteristics have to be investigated by indirect means, usually by measuring the associated ion fluxes across the membrane. The importance of these membrane proteins for excitability is that they give the membrane the ability to pump ions into or out of the cell and to exhibit selective ion permeabilities.

2.4 Membrane Proteins as Ion Pumps

There are a number of membrane proteins which use chemical energy to transport ions across the membrane and which are essential components of any excitable membrane. The ion movement is against the *electrochemical gradient* and is termed *active* transport, as distinct from *passive* transport down a gradient. The concept of an electrochemical gradient will be discussed in the next chapter and it is sufficient to note here that movement of an ion up this gradient requires an investment of energy, just as does the movement of a weight up a gravitational gradient or hill. Moving an ion from a region of low concentration to a region of high concentration, for example, is up a gradient, and the passive flow would be in the reverse direction. Membrane proteins which achieve active transport are termed *ion pumps*, and the best known example of such a pump transports sodium ions out of the cell and potassium ions into it. This sodium pump is found in all cell membranes and is schematically illustrated in Figure 2.6. It produces a coupled movement of sodium out of the cell (efflux) and potassium into the cell (influx), and is driven by chemical energy obtained by the hydrolysis of adenosine triphosphate (ATP). The term 'coupled' indicates that if there is no potassium outside the cell available for pumping in, then the pump will not operate and sodium cannot be pumped out.

The sodium pump creates and maintains ionic gradients across the cell membrane such that sodium is at high concentration outside the cell and low inside while potassium ions are distributed in the opposite fashion. Many cellular functions require the activity of the sodium pump, and the property of excitability is completely dependent on the ion concentration gradients created both by this pump and by other pumps of a similar nature.

22 *Membrane Structure and Properties*

Figure 2.6: The Sodium Pump

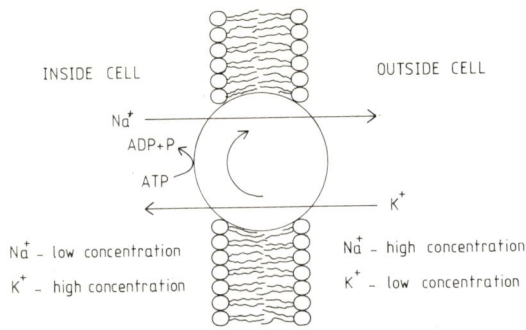

2.5 Membrane Proteins and Passive Ionic Permeability

Excitable membranes possess selective passive ionic permeabilities of two kinds each of which is due to specialised membrane proteins. The first type is a constant ionic permeability which is not affected by physiological stimuli and is termed the resting permeability. It is a common feature of all cell membranes and is primarily a permeability to potassium ions, but sometimes to chloride ions as well, with a much lower sodium ion permeability. The second type of ion permeability is not constant but may change rapidly by orders of magnitude due to the action of a stimulus on the appropriate membrane protein. The permeability mechanism is said to be 'gated' by the stimulus and it is this type of permeability which is characteristic of excitable membranes.

There are in principle two mechanisms by which a protein can confer a selective ion permeability on the cell membrane and these are shown in Figure 2.7. It can act as a *carrier* which binds an ion at one interface, ferries it by diffusion across the bilayer interior and releases it at the other side. Alternatively the protein can span the bilayer and form a pore structure through which the ion can pass from one aqueous phase to the other. The rate at which a carrier protein can move ions across the membrane is severely limited by the slow diffusion rate of such a large molecule in the membrane interior, but a pore protein can transport ions without itself moving and can therefore theoretically achieve a far greater rate of ion flow per protein molecule. A protein known as gramicidin A forms pore structures in artifical bilayers and induces high ionic permeabilities, while valinomycin is a substance which increases bilayer potassium ion permeability by operating as a carrier. As expected, a single molecule of gramicidin can produce far higher rates of ion flow than a single molecule of valinomycin and this is expressed by saying that gramicidin has a higher *unit conductance*.

Figure 2.7: The Carrier and Pore Mechanisms by which a Protein May Allow Ions to Cross the Bilayer.

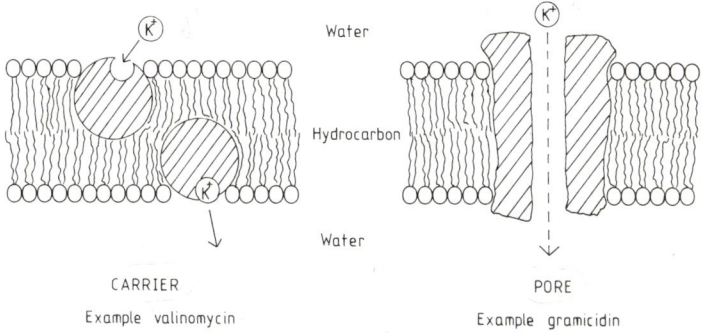

The structures which underlie the resting ion permeability of cell membranes are poorly understood and it is not known whether they operate by a pore or a carrier mechanism. On the other hand the rapidly-gated ion permeability proteins of the excitable membrane (in those examples most studied) have been shown to have very high unit conductances. The findings indicate that their unit conductances are close to that of gramicidin and too high to be explained on the basis of a carrier mechanism. These excitable proteins are therefore believed to be pore structures which span the membrane with an associated gating mechanism which allows them to be turned on and off. The passive ion permeability of an excitable membrane is illustrated in Figure 2.8, in which the flux of ions across the membrane is shown to have two components; one which is always present and another which can be turned on or off by the operation of a gate. The permeability achieved by opening the gated pore structures is far higher than the resting permeability and therefore the gates can, to a large extent, control the ion permeability of the membrane.

Little is known at the molecular level about the way that these gating structures operate, but there is clearly no need for a mechanical blockade of the pore. One possible mechanism is illustrated on the right of Figure 2.8. Here the gating structure is charged and can operate by moving so that the ions are prevented electrostatically from entering the pore. This type of gate is well suited to responding to changes in the voltage across the membrane by opening and closing pores, and such behaviour is found, for example, in nerve membranes. Other gating mechanisms appear to be sensitive to chemical binding, mechanical deformation and perhaps even to light. The coupling of such stimuli to the opening and closing of these permeability mechanisms is fundamental to excitability.

Figure 2.8: Schematic Examples of Proteins Giving the Cell Membrane Constant and Variable Components of Ionic Permeability

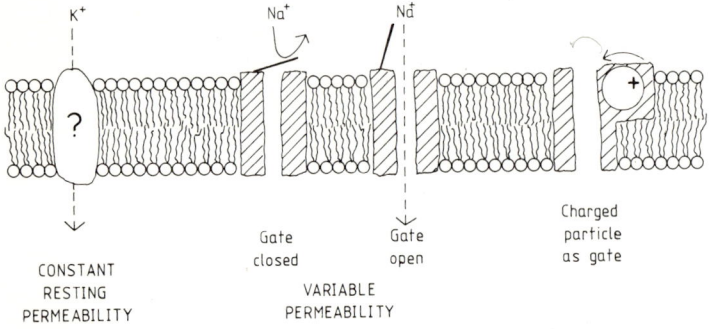

2.6 Summary

The essential components of the excitable membrane and their basic properties have been introduced in this chapter. The bulk of the membrane area is lipid bilayer which is impermeable to ions, and embedded in this matrix are three distinct types of membrane protein. There are membrane proteins which function as ion pumps and which use chemical energy to transport ions up electrical and chemical gradients. There are other proteins which induce a constant resting permeability. Finally and most importantly there are gated permeability channels which are sensitive to a particular stimulus and which are believed to form an ion-conducting pore across the membrane. This type of channel allows the stimulus to control membrane permeability. Other types of protein are present in excitable membranes and may in particular instances be a specialised part of the excitable mechanism, but these three types are the most characteristic and best understood. The next chapter examines how these components function together to produce the property of excitability.

Further Reading

Aveyard, R. and Haydon, D.A. (1973), *Introduction to the Principles of Surface Chemistry.* Cambridge University Press, Cambridge.

Bretscher, M.S. (1973), Membrane structure: some general principles. *Science 181,* 622.

Haydon, D.A. (1975), Functions of the lipid in bilayer ion permeability. *Ann. N.Y. Acad. Sci. 264,* 2.

Haydon, D.A. and Hladky, S.B. (1972), Ion transport across thin lipid membranes: a critical discussion of mechanisms in selected systems. *Q. Rev. Biophys. 5,* 187.

Henderson, R. and Unwin, P.N.T. (1975), Three dimensional model of purple membrane obtained by electron microscopy. *Nature 257,* 28.

Hladky, S.B., Gordon, L.G.M. and Haydon, D.A. (1974), Molecular processes in membranes. *Ann. Rev. Phys. Chem. 25,* 11.

Levine, Y.K. (1972), Physical studies of membrane structure. *Progr. in Biophys. and Molec. Biol. 24,* 1. (X-ray data.)

Singer, S.J. and Nicholson, G.L. (1972), The fluid mosaic model of the structure of cell membranes. *Science 175,* 720.

Stevens, C.F. and Tsien, R.W. (1979), *Membrane Transport Processes, Vol. 3, Ion Permeation through Membrane Channels.* Raven Press, New York.

Tanford, C. (1980), *The Hydrophobic Effect.* 2nd edn, John Wiley and Sons, New York.

Taylor, R.E. (1974), Excitable membranes. *Ann. Rev. Phys. Chem. 25,* 387.

3 THE BASIS OF EXCITABILITY

3.1 Electrical and Chemical Energy

In order to understand the ion flows which take place across excitable membranes it is necessary to have a clear idea of what is meant by the term *electrochemical gradient*. The object of this chapter is to explain this term and then to show how the membrane creates such gradients and uses them to produce the property of excitability.

Figure 3.1: Hypothetical Cell with Electrical Gradient Favouring Potassium Ion Influx.

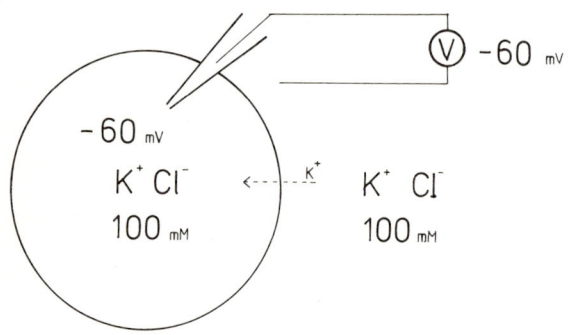

The electrochemical gradient across the cell membrane for an ionic species is a measure of the force driving passive flux of that ion into or out of the cell. As the name suggests it comprises two components — one electrical and the other chemical. The electrical component is simply the electrical potential difference between the inside and outside of the cell, and can be measured by a micropipette electrode as shown in Figure 3.1. Here the internal and external electrodes are connected to a voltmeter and by convention the voltage is expressed as that inside the cell with respect to the outside medium. Like charges repel and unlike charges attract, so this membrane voltage of -60 millivolts (mV) represents a force pulling positive ions (cations) into the cell and pushing negative ions (anions) out. This is the electrical component of the electrochemical gradient for each ion. To quantify this component and to allow comparison with the chemical component we use a quantity called the *free energy* (G) of the ion. The statement that 'an electrochemical force is pushing an ion out of the cell' may otherwise be expressed by saying that by leaving the cell that ion will decrease its free energy. For an ion of valency z in a cell of membrane voltage V the electrical component of the free energy change is given by:

$$\Delta G_{electrical} = zFV \tag{3.1}$$

The symbol ΔG is used to express a change in free energy and F is Faraday's constant of about 96,000 coulombs per mole. Consider the case of the potassium ion (K^+) in Figure 3.1 which is a univalent cation (z = +1):

$$\begin{aligned}\Delta G_{K\ electrical} &= zFV \\ &= 96,000 \times -60 \times 10^{-3} \text{ Joules/mole} \\ &= -5.8 \text{ kJ/mole}\end{aligned}$$

The electrical driving force is therefore now expressed as an energy change per mole of ions. The negative sign indicates that free energy decreases as K^+ moves into the cell and therefore that influx is favoured. For an anion such as chloride (Cl^-) this electrical driving force would be in the opposite direction and ΔG positive.

The chemical component of the electrochemical gradient is simply the chemical concentration gradient for an ion between the inside of the cell (intracellular fluid) and the outside of the cell (extracellular fluid). Ions tend to diffuse down their chemical gradients from regions of high concentration to regions of low concentration. Consider the example of the cell in Figure 3.2 where the membrane voltage is zero but there is a ten-fold concentration difference between K^+ inside and outside the cell. The chemical gradient will tend to cause K^+ efflux and again this can be quantified using the free energy change involved. For an ion in dilute solution the chemical component of free energy change on passing from a region of concentration C_o to a region of concentration C_i is given by:

$$\Delta G_{chemical} = 2.303\ RT \log_{10}\frac{(C_i)}{(C_o)} \tag{3.2}$$

R is the gas constant of 8.3 $JK^{-1}\ mol^{-1}$ and T is the absolute temperature. In this example of a ten-fold chemical gradient the free energy change is

$$\Delta G_{K^+} \text{ chemical} = 2.303 \times 8.3 \times 300$$
$$= +5.8 \text{ kJ/mol}$$

The positive sign indicates that movement out of the cell is favoured.

Figure 3.2: Hypothetical Cell with Chemical Gradient Favouring Potassium Ion Efflux

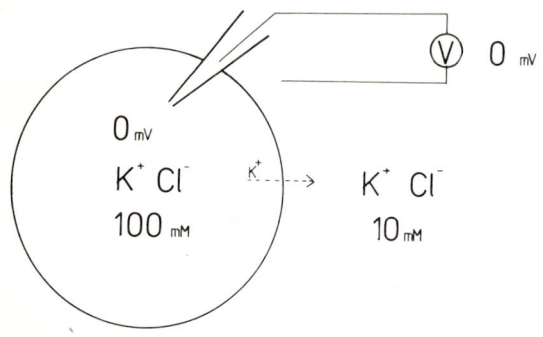

At this point the value of expressing these electrical and chemical driving forces in terms of changes in free energy might seem rather obscure. Indeed we might simply have said that in the first example a 60 millivolt electrical gradient favours K^+ influx, while in the second example a ten-fold concentration gradient favours K^+ efflux. The importance of the free energy expressions is that they allow us to *combine* electrical and chemical gradients to give a single measure of the driving force on the ion. This single measure is the electrochemical gradient and is the sum of the electrical and chemical components of the free energy change.

$$\Delta G_{electrochemical} = \Delta G_{electrical} + \Delta G_{chemical}$$
$$= zFV + 2.303 \, RT \log_{10} \frac{(C_i)}{(C_o)} \quad (3.3)$$

Using this equation, the driving force on an ion can be calculated when both electrical and chemical forces are present at the same time. A positive value for ΔG indicates that the ion will tend to leave the cell and a negative value indicates a force favouring influx.

3.2 The Nernst Equation

When the electrochemical gradient for an ion is zero that ion is said to be in electrochemical *equilibrium* across the membrane. At equilibrium there is no driving force for ion flow and the passive efflux of ions will be exactly balanced by the passive influx. There are still ion movements across the membrane but there is no *net* flux. Clearly this equilibrium is present if the electrical and chemical driving forces are both zero as when $V = 0$ and $C_i = C_o$. However, equilibrium is also present if the electrical and chemical driving forces are *equal* and *opposite*.

$$\text{When } \Delta G_{electrochemical} = 0 \tag{3.4}$$

$$\text{From (3.3) } \Delta G_{electrical} = -\Delta G_{chemical}$$

$$zFV = -2.303 \, RT \, \log_{10}\frac{(C_i)}{(C_o)}$$

$$V = 2.303 \, \frac{RT}{zF} \, \log_{10}\frac{(C_o)}{(C_i)} \tag{3.5}$$

Equation (3.5) is known as the Nernst equation and it describes the conditions for electrochemical equilibrium for an ion whose extracellular and intracellular concentrations are C_o and C_i respectively. The value of membrane voltage V which satisfies this equation is known as the Nernst potential or equilibrium potential for that ion and may be written V_k, for example, in the case of the potassium ion. The Nernst equation contains the expression 2.303 RT/F which at physiological temperatures is approximately equal to 60 millivolts. Thus for an univalent ion a ten-fold concentration gradient can be roughly balanced by a 60 mV electrical gradient in the opposite direction. Figure 3.3 illustrates an example in which the membrane voltage is -60 mV and the internal and external K^+ concentrations 100 and 10 mM respectively. Applying the Nernst equation:

$$V_k = 2.303 \, \frac{RT}{ZF} \, \log_{10}\frac{(C_o)}{(C_i)} \tag{3.6}$$

$$= -60 \, mV$$

we can see that the measured membrane potential V is equal to the equilibrium potential V_k and therefore the potassium ion is at electrochemical equilibrium across the cell membrane. This example

30 *The Basis of Excitability*

is a combination of Figures 3.1 and 3.2 and it is now clear why the driving forces in those examples were equal and opposite.

Figure 3.3: The Potassium Ion is at Electrochemical Equilibrium due to Equal and Opposite Electrical and Chemical Forces.

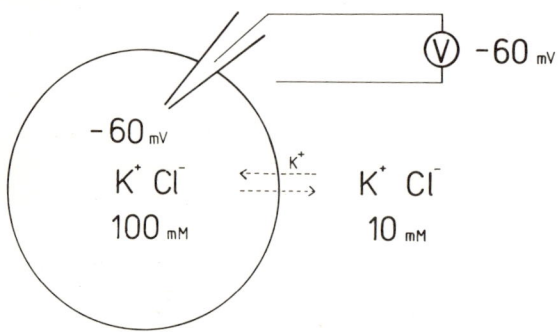

The value of the Nernst equation is that it allows a calculation of the electrochemical driving force on the ion. First the external and internal ion concentrations are measured and the Nernst equation applied to find the equilibrium potential (denoted V_k for potassium, V_{Na} for sodium and so on). Then considering potassium ions the measured membrane voltage V is compared to V_k using the quantity $(V - V_k)$.

$(V - V_k) = 0$ Electrochemical equilibrium
$(V - V_k) > 0$ Gradient Favouring K^+ efflux
$(V - V_k) < 0$ Gradient favouring K^+ influx

The importance of the sign of $(V - V_k)$ in predicting the direction of passive potassium ion flux can be deduced from the definition of electrochemical gradient (equation 3.3) and the Nernst equation (3.6)

$$\Delta G_{k\ electrochemical} = zFV + 2.303\ RT \log \frac{(C_i)}{(C_o)} \quad (3.3)$$
$$= zFV - zFV_k \text{ (from 3.6)}$$
$$= zF(V - V_k) \quad (3.7)$$

3.3 The Active Transport of Ions

If the net flux of ions across the cell membrane is against the direction of passive flow then they are moving up an electrochemical gradient. Figure 3.4 shows a cell in which both electrical and chemical forces favour sodium ion influx. The rates of sodium ion movement out of the cell (J_1) and into the cell (J_2) can be measured using radioactive tracers, and the net flux is the difference between J_1 and J_2. If efflux exceeds influx then the ions are moving up the gradient and *active transport* is occurring. This requires an energy supply, which is usually supplied by adenosine triphosphate (ATP), and the transport is performed by the ion-pump proteins briefly discussed in Chapter 2. Often it is found that J_1 and J_2 are equal and the intracellular sodium ion concentration is constant. (If the membrane was completely impermeable to sodium ions ($J_1 = J_2 = 0$) then this could be explained without active transport but as long as any passive leak is occurring it can only be balanced by ion pumping.)

Figure 3.4: Active and Passive Fluxes of Sodium and Calcium Ions. The total efflux of sodium ions (J_1) is the sum of active and passive components.

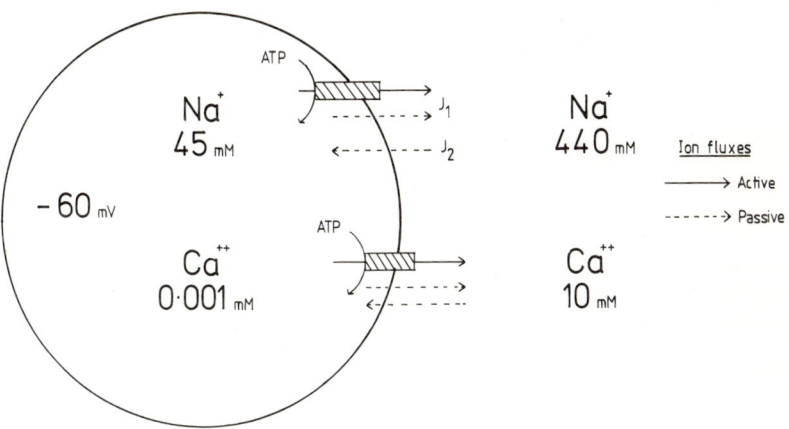

Ion pumps move ions up electrochemical gradients and serve to create and maintain those gradients, while passive fluxes tend to dissipate them. In the living cell ion concentrations remain constant because passive fluxes are balanced by active pumping, as illustrated in Figure 3.4 for sodium and calcium ions. It is simplest to consider the ion pumps as the creators of chemical concentration gradients across the cell membrane. This is not strictly correct as some pumps also drive an electric current and therefore contribute directly to membrane voltage. However in excitable cells membrane voltage is primarily determined by the ionic concentration gradients and the *passive* permeability of the membrane to ions.

3.4 Resting Membrane Ion Permeability and Membrane Voltage

Figure 3.5: The Influence of Selective Membrane Ionic Conductance on Membrane Voltage

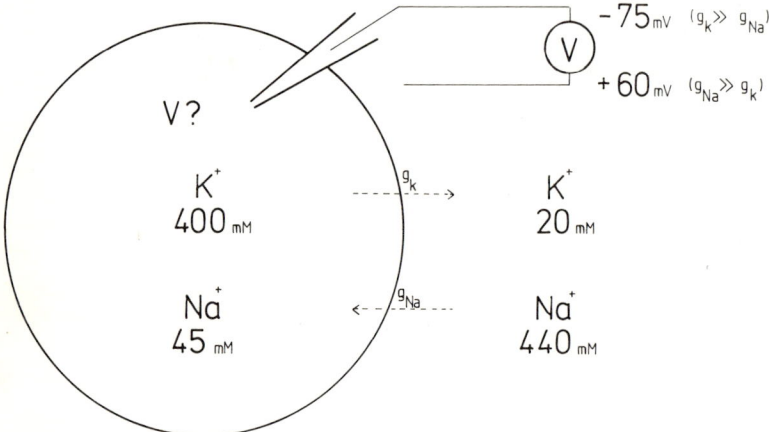

Consider the example of Figure 3.5 in which the ion pumps have created an asymmetrical distribution of sodium and potassium ions. Additional substances are present both inside and outside the cell so that the total solute concentration is equal inside and outside, and so anion and cation concentrations are equal. For present purposes only the two ionic species shown are important. Initially suppose that the membrane voltage is zero and so there is a tendency for potassium ions to flow out of the cell and for sodium ions to flow in. Whether the membrane voltage becomes positive or negative depends on the relative permeability of the membrane to potassium and sodium ions. If the membrane is permeable to potassium but not to sodium then K^+ will flow out of the cell and the membrane voltage will become negative as the cell loses positive charges. The potassium ion efflux will continue until the negative membrane voltage pulling potassium ions back into the cell is equivalent to the chemical concentration gradient pushing them out — in other words until the Nernst equation holds for potassium ions. Usually the amount of ion flux required to produce this change in potential is small and does not change the chemical gradients. Therefore the membrane voltage attained is given by

$$V = V_k = 2.303 \frac{RT}{zF} \log_{10} \frac{(20)}{(400)}$$
$$= -75 \text{ mV}$$

By exhibiting a selective permeability to potassium the membrane has achieved two important conditions using the pre-existing concentration gradients: (1) a resting membrane potential with the inside of the cell negative; (2) an increased electrochemical gradient favouring sodium ion entry.

Returning to the initial conditions of Figure 3.5; if the membrane were permeable to sodium and not to potassium ions then an inward flux of sodium would occur. In this case the membrane voltage would become positive and approach the Nernst potential for sodium ions V_{Na}:

$$V_{Na} = 2.303 \frac{RT}{zF} \log_{10} \frac{(440)}{(45)} = +60 \text{ mV}$$

It appears that by changing its passive permeability the membrane can determine its own voltage between the limits of -75 and $+60$ mV or V_k and V_{Na}. In general terms, by exhibiting a selective permeability for an ion the membrane voltage is caused to approach the Nernst potential for that ion. The passive flux of an ion is always in such a direction as to move the membrane voltage towards that ion's Nernst potential. If the membrane is significantly permeable to two ionic species at once then the membrane voltage attains a value between their Nernst potentials.

It ought now to be clear that the passive flux of an ionic species across the cell membrane is a function of the electrochemical driving force on that ion and of the membrane permeability to that ion. It is useful to have an expression linking these quantities, and the simplest method is to assume that currents across the membrane obey Ohm's law: $V = IR$ or $I = V/R$. The current I_x across the membrane (due to an ion X) is related to its driving force $(V - V_X)$ by the membrane conductance to that ion g_X:

$$I_X = g_X (V - V_X) \tag{3.8}$$

Conductance is the reciprocal of resistance and has the units of reciprocal ohms or siemens (S). Now we can estimate the membrane voltage attained by passive ion flows in Figure 3.5 when both sodium and potassium ion movements occur. This stable membrane voltage will be that at which sodium and potassium fluxes balance and total membrane current I_m is zero.

$$I_m = I_{Na} + I_k$$
$$= g_{Na}(V - V_{Na}) + g_k(V - V_k) = 0$$

Therefore
$$V = \frac{g_k V_k + g_{Na} V_{Na}}{g_k + g_{Na}} \tag{3.9}$$

If, for example sodium and potassium ion conductances are equal ($g_{Na} = g_k$) then membrane voltage will approach ½ $(V_{Na} + V_k)$ or -7.5 mV. This situation is in equilibrium in terms of electric current flows only as the opposing fluxes of sodium and potassium ions are still causing chemical changes.

The resting excitable cell membrane exhibits a constant selective ionic permeability to potassium ions far greater than its sodium ion permeability. This is due to the resting permeability mechanisms mentioned in Chapter 2 which ensure that g_k is much larger than g_{Na}. The membrane voltage therefore lies close to V_k and the cell is said to have a *resting potential* of about -70 mV. This has important consequences for the electrochemical gradient experienced by the sodium ion, which is now very large. The flux of sodium ions across the membrane is small only because of the low membrane conductance to sodium.

$$I_{Na} = g_{Na} \times (V - V_{Na})$$
low   very   high
         low $+130$ mV

The electrochemical gradient favouring entry of sodium ions into the cell is a considerable energy source. A gravitational equivalent would be the energy stored in the lake of water held upstream by a dam. It is this type of electrochemical energy source which is used by excitable cells to amplify and transmit signals. These electrochemical gradients are utilised by the gated ion permeability channels of the excitable membrane.

3.5 Gated Ion Channels and Excitability

In the typical excitable cell in its resting state both sodium and calcium ions experience a large inward electrochemical driving force, but only enter slowly as the membrane conductance to these ions is very low. These gradients are illustrated in Figure 3.6 together with ion channels (A and B) permeable to calcium and sodium, which are gated by specific stimuli. If a sodium channel of the pore type (A) is opened by a stimulus, then a large current of sodium ions will pass into the cell driven by the pre-existing electrochemical gradient and not by the energy of the stimulus. The conductance of a single pore is such that many thousands of ions may enter the cell in one millisecond due to the operation of its gate. This functional unit is therefore a means of signal amplification and may be termed an *excitable* ion channel.

The ion flows which occur through these excitable channels may in principle be sensed by the cell in one of two ways: (1) the current associated with ion flow causes a change in membrane voltage; (2) the ion flow changes the intracellular concentration of that ion. In distinguishing these possibilities it is important to realise that changes in the absolute value of ion concentration are not so important as changes in log (concentration). Sodium, potassium and chloride ions have intracellular concentrations in the range 10 to 500 mM, while intracellular calcium ion concentration is very low at about 0.001 mM. Consider the effect of equal flows of calcium and sodium ions into the cell, sufficient to raise the concentration of each ion by 0.01 mM. The internal calcium ion concentration is raised by a factor of eleven to 0.011 mM, but there is a negligible effect on internal sodium ion concentration as it changes by only a fraction of one per cent. The very low level of intracellular calcium ions means, therefore, that flows of calcium are usually detected by the resultant changes in ion concentration. Sodium, potassium and chloride ion flows are usually sensed by their effect on membrane voltage.

Figure 3.6: Examples of Gated Ion Channels for Sodium, Calcium and Potassium Ions

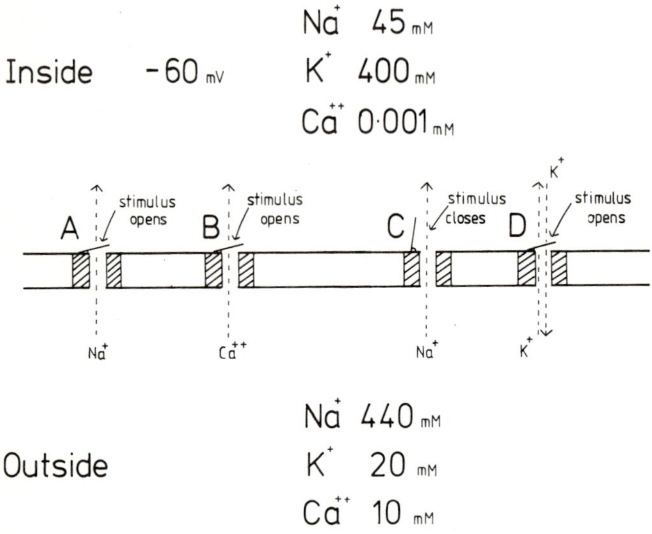

Inside -60 mV

Na^+ 45 mM
K^+ 400 mM
Ca^{++} 0.001 mM

A stimulus opens — Na^+
B stimulus opens — Ca^{++}
C stimulus closes — Na^+
D stimulus opens — K^+

Outside

Na^+ 440 mM
K^+ 20 mM
Ca^{++} 10 mM

In its simplest form the excitable mechanism involves the opening of membrane channels to an ion far from electrochemical equilibrium by a stimulus, such as the binding of a specific chemical substance. Variations on this theme exist and two of these are illustrated in Figure 3.6. Ion channel C is open in the resting state and conducts an inward current of sodium ions. The stimulus closes the channel and therefore stops the current. Again an amplification has occurred as the stimulus has in effect produced a current in the opposite direction to the original ion flow, and membrane voltage will be significantly affected. This type of channel appears to be important in the visual system (Chapter 6). Ion channel D is permeable to potassium ions which in the resting state are already at electrochemical equilibrium across the membrane. Opening of this channel will, by itself, produce no change in the electrical or chemical state of the cell. It will have an effect, however, if any other mechanism has changed the membrane voltage from its resting level and will always tend to restore the resting conditions. Channels of this type allow two stimuli to have opposing actions on the cell, and they can speed up the return to a resting state.

One final concept which may be introduced at this stage is that of a *regenerative mechanism.* Consider an example where channels allowing sodium ions to enter the cell are opened by an increase in membrane voltage. The sodium ion flow will itself cause a further increase in membrane voltage and tend to open more channels. This is a regenerative cycle as illustrated in Figure 3.7, as the stimulus is self-reinforcing. Positive feedback of this type allows a greater amplification of the stimulus than each ion channel could achieve acting independently. The stimulus must reach a certain *threshold* strength before the regenerative mechanism can overcome the natural decay of the signal. The production of a regenerative response to stimuli above a critical threshold is found in many excitable tissues, and most notably in nerve cells.

Figure 3.7: A Self-regenerative Cycle which Amplifies changes in Membrane Voltage

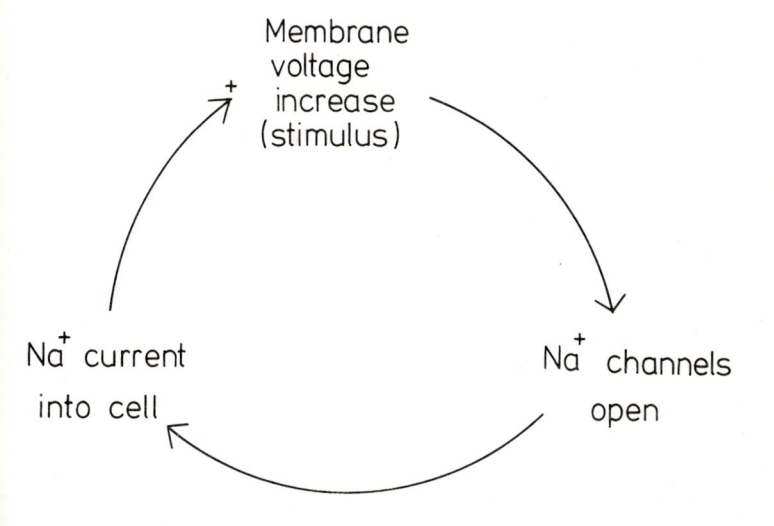

3.6 Summary

In this chapter the *electrochemical gradient* has been introduced and defined. It is the force which causes passive ion flows across the cell membrane, and it can be divided into electrical and chemical components. In certain conditions these two components may oppose and cancel each other out, in which case the ion in question is at electrochemical equilibrium across the membrane. At equilibrium the relationship between the ion concentrations inside and outside the cell and membrane voltage is given by the *Nernst equation*. If the ion concentrations are known then this equation can be used to calculate the *equilibrium potential* — the membrane voltage at which that ion would be at equilibrium. For an ion X the equilibrium potential is written V_X. The difference between the observed membrane voltage (V) and this equilibrium potential is a measure of the direction and magnitude of the electrochemical gradient for the ion.

Ion pumps in the excitable membrane move ions up electrochemical gradients and serve to create and maintain ion concentration gradients. Once these concentration differences are established the membrane voltage is primarily determined by the selective passive membrane peremability to the ions present. The ease with which an ion X can cross the membrane may be quantified using the membrane conductance to that ion g_X. In the resting state there is a higher potassium conductance (g_k) than sodium or calcium conductance and the resting membrane voltage is therefore close to V_k. The membrane conductance to each ion may be altered rapidly and reversibly by means of gated ion channels. The stimuli which open and close these channels can thereby cause significant ion flows across the membrane. These flows are driven by the energy stored in the existing electrochemical gradients and represent a substantial amplification of the stimulus. The importance of this ability to undergo rapid and reversible changes in membrane ion conductance will become clear as individual examples of the excitable tissues are discussed in the following chapters.

Further Reading

Cole, K.S. (1968), *Membranes, Ions and Impulses.* University of California Press, Berkeley.

Hodgkin, A.L. (1964), *The Conduction of the Nervous Impulse.* Liverpool University Press, Liverpool.

Katz, B. (1966), *Nerve, Muscle and Synapse.* McGraw Hill, New York. (Chs 1-4.)

Moore, W.J. (1972), *Physical Chemistry.* 5th edn, Longmans, London. (Chs 10-12.)

Part Two

THE COMMUNICATION OF INFORMATION

4 THE NERVOUS IMPULSE

4.1 Introduction

The nervous impulse is the only mechanism which can achieve a rapid communication of information between distant parts of a large organism. The structure of a typical nerve cell is illustrated in Figure 4.1. The long cylindrical process of the cell, known as an axon, is responsible for carrying impulses from the cell body to its terminal which may be many feet away. These impulses are waves of electrical disturbance which are propagated down the axon by virtue of its specialised membrane and this chapter is concerned with an explanation of this excitable phenomenon.

Nerve axons vary enormously in size but the largest known have been found in the squid and these can be as much as 1 mm in diameter. A great deal of what is now known about the nervous impulse has been discovered using these squid 'giant axons' because an electrode can be inserted *into* the axon itself and measurements made of the internal electrical potential. Knowledge of this intracellular potential is vital to an understanding of the nerve impulse and so we begin with a description of the impulse in squid nerve preparations.

Figure 4.1: Schematic Diagram of a Nerve Cell

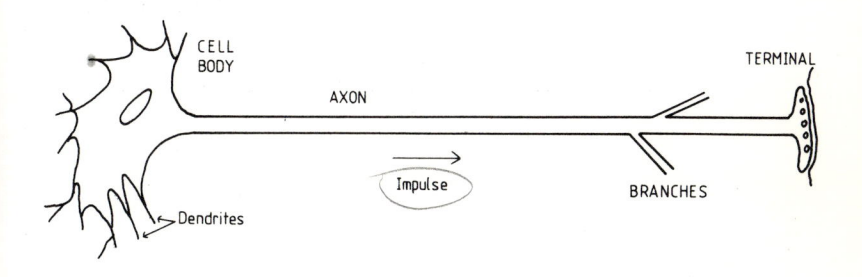

4.2 The Squid Nerve Action Potential

Squid giant axons can be dissected out of recently killed specimens and will continue to conduct impulses for many hours. Figure 4.2 is a diagram of such an axion showing how the internal electrical potential can be measured by inserting an electrode along the fibre. The potential difference between the inside and outside of the axon is known as the membrane potential or membrane voltage and is expressed as the potential inside with respect to the outside. At rest this membrane voltage (V) is typically −70 millivolts (mV). the passage of an impulse causes a characteristic change in membrane potential known as the *action potential*. By passing current through the stimulating electrodes shown in Figure 4.2 an impulse can be initiated which will travel along the axon past the end of the recording electrode. The potential measured by the recording electrode as this occurs is shown in Figure 4.3 which is a graph of membrane voltage against time. The transient change in membrane potential from −70 mV to +40 mV and back again is the action potential and is the most important manifestation of the nervous impulse: indeed the terms action potential and nervous impulse are almost interchangeable.

Figure 4.2: Measurement of Membrane Voltage in the Squid Giant Axon

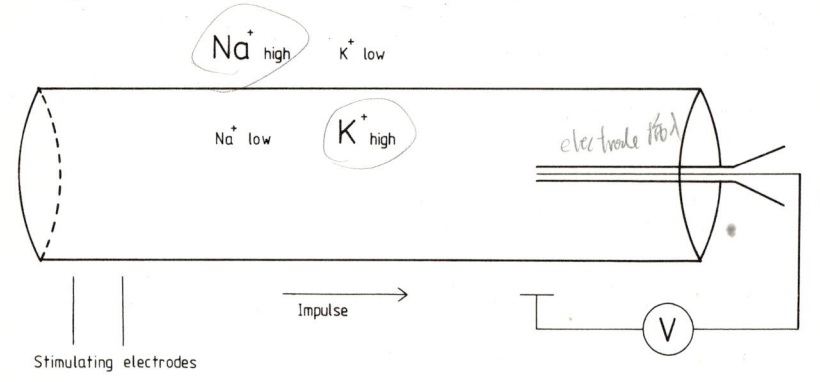

A number of features of the action potential can be deduced from measurements of the type illustrated in Figure 4.3. The application of the stimulating current produces an immediate effect at the recording electrode due to direct current flow. This *stimulus artefact* is marked SA in the figure. The delay between the stimulus artefact and action potential represents the time taken for the impulse to travel along the axon to the recording electrode. The velocity of propagation can be measured by placing the stimulating and recording electrodes at different separations and noting the change in delay. In the squid axon this velocity is close to 20 metres per second. Furthermore this velocity is constant along the axon so the action potential neither accelerates nor slows as it moves along. The form of the action potential is also of interest as it actually involves a fast reversal of the negative resting membrane potential to positive values, and a slower return to the resting level in about 2 milliseconds (ms). If the strength of the stimulating current is reduced there comes a point when no action potential is produced. There is no halfway stage at which an action potential of reduced size or velocity is produced; in other words the action potential is an 'all-or-nothing' phenomenon. Associated with this is the observation that there is a minimum stimulus strength which will just produce an action potential, while any weaker stimulus will not. This is the *threshold stimulus*. As the action potential passes along the axon it does not change in shape or height and thus the propagation is non-decremental.

Figure 4.3: The Action Potential

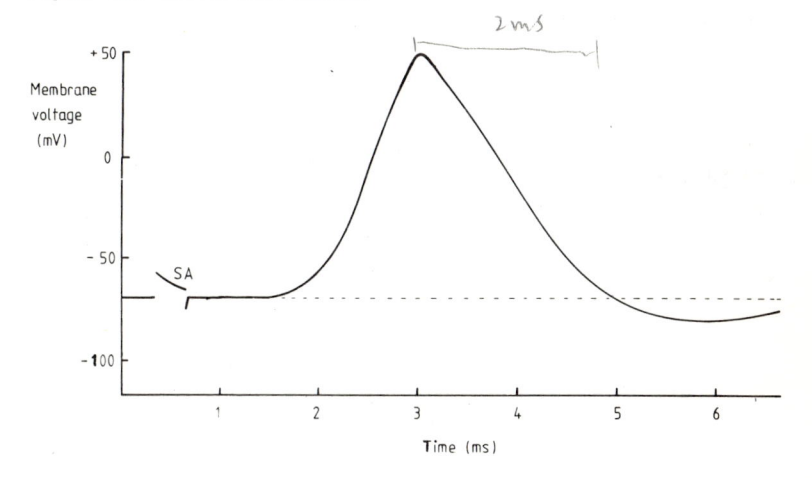

If a second stimulating current is passed immediately after the first then it is more difficult, or even impossible, to elicit a second action potential. This is the phenomenon of a *refractory period*. The threshold stimulus strength needed to initiate a second action potential is shown in Figure 4.4 as a function of the time since the first stimulus. There is an absolute refractory period (ARP) immediately after the first stimulus lasting for approximately the duration of the action potential, during which it is impossible to initiate a new impulse. There follows a period of increased threshold known as the relative refractory period. The ARP puts an upper limit on the number of action potentials which can travel down the axon per unit time, this number being the maximum impulse frequency.

Figure 4.4: The Refractory Period Following an Action Potential

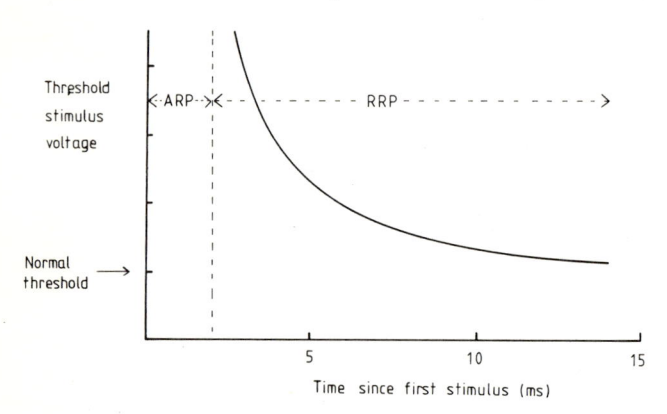

The following characteristics of the action potential have therefore been outlined. It is a transient reversal of membrane potential which propagates along the axon at a constant velocity in a non-decremental manner. It is an all-or-nothing phenomenon which is initiated by stimuli above a threshold strength, and during its passage the axon is absolutely refractory. These characteristics are found both in squid nerves and in those mammalian axons which have been studied.

4.3 Ion Concentration Gradients and the Sodium Pump

A further advantage of the large size of the squid axon is that it is relatively easy to measure the ionic concentrations in the intracellular fluid or axoplasm. For conduction of the nervous impulse the important ions are sodium and potassium. The intracellular concentration of sodium is approximately 45 mM compared with an external concentration of 440 mM. Potassium ions have an internal concentration of

400 mM and an external concentration of 20 mM. The axoplasm therefore has an accumulation of potassium ions and is depleted in sodium. These ionic concentration gradients are vital to the production of an action potential; they are created and maintained by the sodium pump introduced in Chapter 2 (page 21).

Figure 4.5: Procedures which Alter the Rate at Which Sodium Ions Are Pumped out of the Squid Axon

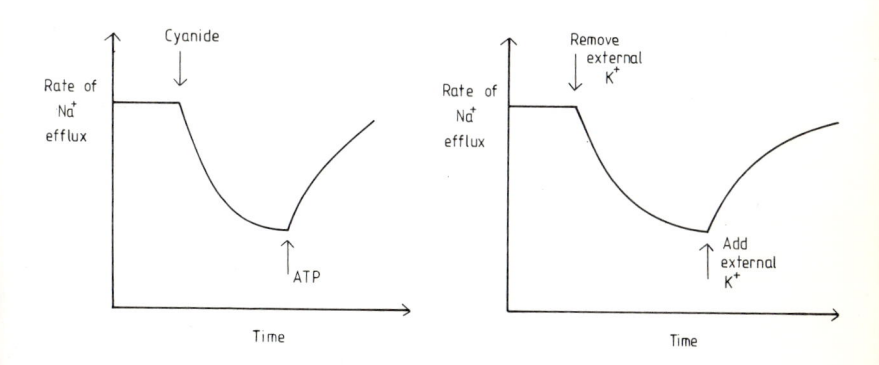

The pump can be investigated in the squid axon by using radioactive tracers such as $^{24}Na^+$ and two sample experiments are illustrated in Figure 4.5. The axon is loaded with $^{24}Na^+$ and then superfused with an external medium containing no radioactivity, so that an estimation can be made of the rate of loss of internal sodium. This rate is found to be reduced by the addition of cyanide or other metabolic inhibitors and then increased again by the addition of adenosine triphosphate (ATP). The loss of sodium ions is also inhibited by removal of potassium ions from the external medium. Experiments of this type demonstrate that the squid sodium pump operates in a manner similar to the pump in other tissues. The efflux of sodium is coupled to an influx of potassium as was illustrated in Figure 2.6. The drug ouabain blocks the pump by direct binding to the pump protein and this inhibition is not reversed by addition of ATP.

The sodium pump is present in some form in all cell membranes and its activity is vital to many cell functions. In the production of the nervous impulse its role is simply to create the appropriate sodium and potassium ion concentration gradients. Even if the pump is blocked with ouabain the axon will continue to conduct impulses as long as the ion gradients remain. If the gradients are dissipated then excitability is

46 The Nervous Impulse

lost completely. In practice this means that an axon can function for many hours after the pump has been blocked, but in the long term the pump is necessary for excitability.

4.4 The Resting Potential

Given the sodium and potassium ion distribution across the axon membrane, it is now possible to understand why the resting membrane potential is found to be −70 mV. In Chapter 3 it was demonstrated how the membrane potential could be determined by the passive ionic permeability of the membrane if ionic concentration gradients existed. In the extreme case where the membrane is only permeable to one ionic species its potential approaches the Nernst potential for that ion. The squid axon resting potential is close to the Nernst potential for the potassium ion.

$$V_K = \frac{RT}{zF} \log \frac{(K_0^+)}{(K_i^+)} = -75 \text{ mV}. \tag{4.1}$$

The present theory of the resting potential is that it results from a selective permeability to potassium ions in the axon membrane. The evidence for this view is twofold. First, studies with radioactive tracers confirm that potassium ions do diffuse across the axon membrane more easily than sodium ions. Secondly a direct test can be performed by varying the external and internal potassium ion concentrations and measuring the change in resting potential. The results of such studies are illustrated in Figure 4.6 and it is seen that the resting potential continues to lie close to V_K. The graphs show the expected slope of about 60 mV per ten-fold concentration change. The deviations from theory at high negative values of V_K are not fully understood but involve the fact that some small sodium ion permeability is present.

The resting potential is therefore a consequence of (1) the ionic gradients created by the sodium pump and (2) the selective permeability of the resting axon membrane to potassium with respect to sodium ions. An additional factor which may be important in some nerve cells is that the sodium pump has been found capable of transporting more sodium ions out of the cell than it transports potassium in, and such electrogenic pumping would contribute directly to the negative internal potential. The maintenance of a near-normal resting potential is a pre-requisite for nerve excitation. This is because the gated ion channels which produce the action potential are adapted to function with this resting potential. In addition the negative resting potential means that there is a large electrochemical gradient favouring sodium ion entry which is exploited by the active membrane.

Figure 4.6: The Influence of External and Internal Potassium Ion Concentration on the Resting Potential. The slopes of the graphs are consistent with the Nernst equation.

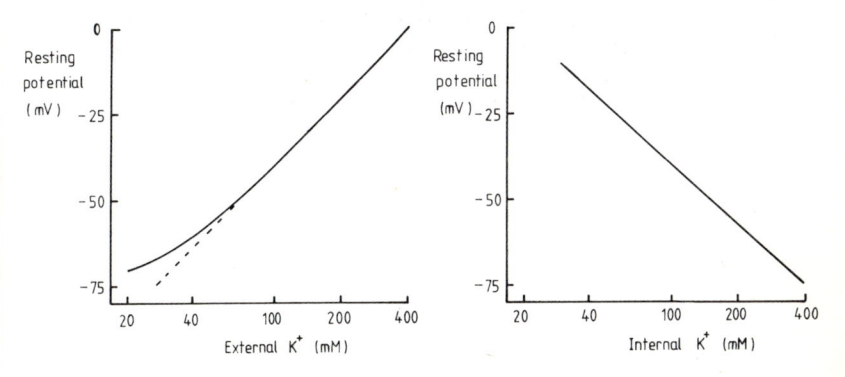

4.5 The Active Membrane

As the action potential passes, the axon membrane in that region is said to be in an active state and this state is characterised by the reversal of membrane voltage from negative to positive values. By convention an increase in membrane potential is called a *depolarisation* and the active membrane undergoes a rapid depolarisation which overshoots the zero membrane potential level to approach +40mV. Following the arguments of Chapter 3 on the possibility of the membrane potential varying between the limits of the Nernst potentials of the ionic species present, it is of interest to know V_{Na} for the axon:

$$V_{Na} = \frac{RT}{zF} \log \frac{(Na_o^+)}{(Na_i^+)} = +60 \text{ mV} \qquad (4.2)$$

This means that it is possible that the action potential results from an increase in the membrane conductance to sodium ions. The simplest evidence that this is the case may be summarised:

(1) If the external concentration of sodium ions is reduced then the action potential becomes smaller in height and always remains below V_{Na}. As sodium ion concentration is further reduced no action potential will occur.

(2) The total membrane electrical conductance during an action potential is vastly increased, as was first demonstrated by Cole and Curtis (1939).

Figure 4.7: A Voltage Clamp Experiment — TTX Blocks the Early Transient Inward (Sodium Ion) Current while TEA Blocks the Maintained Late (Potassium Ion) Current

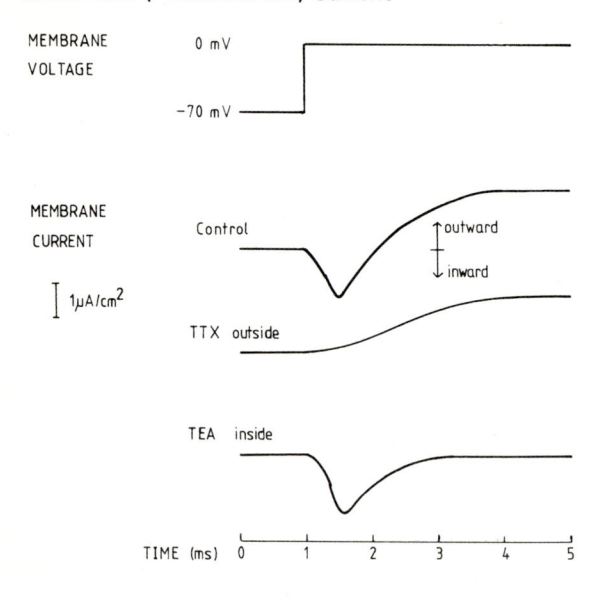

The 'sodium hypothesis' is that an increase in sodium ion conductance (g_{Na}) leads to an influx of sodium ions and therefore an increase in membrane potential towards V_{Na}. Confirmation of this hypothesis was achieved by means of an experimental technique known as the voltage clamp. A full discussion of the method is beyond the scope of the present book, but it is a procedure whereby the membrane voltage is imposed by the experimenter and varied in a step-wise fashion, while a simultaneous recording of membrane current is made. A typical set of experimental records is shown in Figure 4.7. Initially the membrane potential is clamped at −70 mV and it is then rapidly changed in a step up to 0 mV. The membrane current resulting from this step is shown and it reveals an initial transient inward component followed by a maintained outward component. The transient inward current is a result of sodium ion influx while the late outward current is due to potassium ion efflux. Thus, for example, the inward current is reduced by decreasing external sodium ion concentration. The sodium and potassium ion currents are quite separate and can be independently blocked by the substances tetrodotoxin (TTX) and tetraethylammonium (TEA) respectively. TTX blocks the sodium ion current as shown in Figure 4.7 while TEA blocks the potassium ion current.

Figure 4.8: The Sodium and Potassium channels of the Nerve Membrane. The sodium channel inactivates and so behaves as if it has two gates. The sodium pump illustrated on the right is a separate entity.

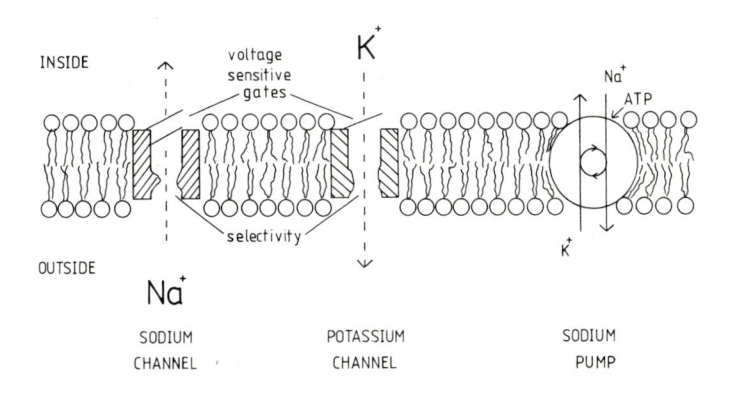

These voltage-clamp currents were first analysed by Hodgkin and Huxley who demonstrated that they could be explained by a model in which the axon membrane contains two types of gated conductance channel, a sodium channel and a potassium channel. These independent channels are illustrated in Figure 4.8. Apart from being selective for either sodium or potassium ions, and having a high unit conductance, the important properties of these channels reside in their gating mechanisms. The opening and closing of these channels are time and voltage-dependent and their fundamental properties are:

(1) The sodium channel is rapidly opened by a depolarisation of the membrane voltage but then slowly turns itself off or

'inactivates' even if the depolarisation is maintained. The open phase is always transient as in the TEA trace of Figure 4.7.

(2) The potassium channel is slowly opened by a depolarisation of the membrane and does not inactivate — at least not on the short time scale of the action potential. There is therefore a late outward potassium current as long as the depolarisation is maintained, as in the TTX trace of Figure 4.7.

Using a mathematical description of this channel model, Hodgkin and Huxley (1952) were able to predict correctly the form, height, conduction velocity and threshold properties of the action potential. This was a remarkable achievement, as the model was derived from voltage-clamp data and yet it successfully represents membrane behaviour when voltage is freely variable. The Hodgkin-Huxley model and equations have therefore become the point from which studies of many excitable membranes begin. It is of interest to consider how it is that the action potential is determined by the properties of the sodium and potassium channels.

Consider first the phenomenon of a threshold stimulus producing an 'all-or-nothing' response. The sodium channel is opened by membrane depolarisation and the current of sodium ions passing through the channel into the axon will cause more depolarisation, thereby opening more channels and further increasing sodium ion influx. This is a self-regenerative phenomenon as discussed in Chapter 3 (page 37). It might appear that any depolarisation, however small, would begin this chain reaction and open all the sodium channels. This is not so because the resting membrane permeability to potassium ions results in an outward flux of potassium which will resist depolarisation. Furthermore sodium channels do not open in significant numbers until the membrane voltage has moved about 20 mV from the resting level, after which point a small change in voltage opens many channels. Thus there exists a threshold depolarisation at which the potassium ion efflux is just balanced by sodium ion entry. Only a stimulus which just exceeds this threshold will initiate an action potential. The self-regenerative nature of sodium channel activation means that once the threshold is exceeded a full activation will always take place so that the action potential is 'all-or-nothing', and is constant in height.

The transient nature of the action potential is the result of two processes both of which are incorporated in the Hodgkin-Huxley model. First the sodium channel inactivates as a result of the positive membrane voltage, and sodium ion conductance falls. Secondly the depolarisation causes a slow or delayed opening of the potassium channels thereby increasing potassium conductance and drawing the membrane voltage back towards V_k and the resting potential. These effects account for the rapid turnaround of membrane voltage at the peak of the action potential and the subsequent repolarising phase. The refractory period is also associated with these two processes. Inactivated sodium channels are not opened again by immediate depolarisation and take a few milliseconds to become functional once more. Increased potassium ion conductance always tends to increase the potassium ion currents which resist any movement of membrane voltage away from the resting level.

4.6 The Propagation of the Impulse

Figure 4.9: The Local Circuit Theory of Impulse Propagation. Note that the graph is of voltage versus distance and so its appearance is back-to-front compared with Figure 4.3.

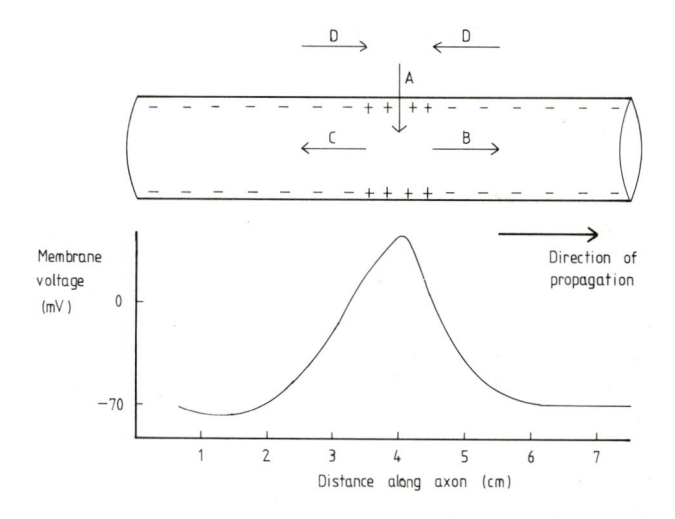

The propagation of the action potential along the axon results from local circuit currents which flow longitudinally along the cylindrical structure both inside and outside the cell. These local currents are illustrated in Figure 4.9 in which the directions of current flow are represented by arrows near the active region of the membrane. The lower section of this figure is a graph of instantaneous potential versus distance along the axon. Current A is the influx of sodium ions at the active membrane. The graph indicates that the internal electric potential is at a maximum at the site of activity. Therefore currents will flow longitudinally inside the axon away from that site as indicated by arrows B and C. Current B moves ahead of the impulse and tends to depolarise the axon membrane in the next segment. This current is a sufficient stimulus to excite the membrane ahead and so propagate the impulse. Current C, which travels back down the axon, does not cause a re-excitation because the membrane behind the impulse is refractory. The circuit of local currents is completed by the arrows marked D which represent longitudinal flow outside the axon. These currents are all carried by ions in the aqueous medium surrounding the nerve membrane.

The simplest evidence for local circuit currents as the method of propagation lies in observations showing that changes in resistance to these currents alter the impulse conduction velocity and can block the axon completely. The resistance to internal current flow is lowest in axons of large diameter and, as expected, these axons exhibit higher conduction velocities. The relationship between diameter (d) and conduction velocity (θ) is found to be:

$$\theta = \text{constant} \times \sqrt{d}$$

This relationship is the one predicted from the electrical properties of the axon assuming local currents as in a cable. It is possible to change the diameter of a single squid axon by canulating the axon cylinder, and literally blowing it up or contracting it by injection or removal of fluid. The fully-expanded axon has the highest conduction velocity. The resistance to the external local currents can also be altered. Placing a metal conductor beside the axon increases θ, while partially drying a segment of the axon decreases θ and can block propagation completely.

For a fibre of given diameter, conduction velocity can be vastly increased by a process known as *myelination* and the structure of a myelinated nerve is shown in Figure 4.10. The myelin sheath is composed of Schwann cells which wrap themselves many times around the axon creating a thick layer of multiple cell membranes. At intervals along the axon the sheath disappears at the nodes of Ranvier. The myelin sheath functions as a high-resistance barrier to current flow which modifies the cable properties of the axon and the pattern of local circuit currents. In a myelinated axon excitation of the membrane occurs only at the nodes of Ranvier. Between these nodes little or no current flows across the axon membrane, as illustrated in Figure 4.10. The local currents from the active node pass longitudinally down the axon to depolarise the next node along, and the myelin prevents leakage of the current out of the axon from the internodal segment. These currents excite the next node and so the impulse is propagated by skipping from node to node in a process known as saltatory conduction (from the Latin, *salto:* I jump). Apart from having a far higher conduction velocity, myelinated nerves require less energy to be expended in ion pumping. The ion flows of the action potential tend to dissipate the ionic concentration gradients created by the sodium pump and in the long term the pump must counter these changes. As excitation occurs only at the nodes of Ranvier in a myelinated nerve, the total ionic flux due to an impulse is far smaller than in an unmyelinated nerve.

Figure 4.10: Saltatory Conduction in a Myelinated Axon

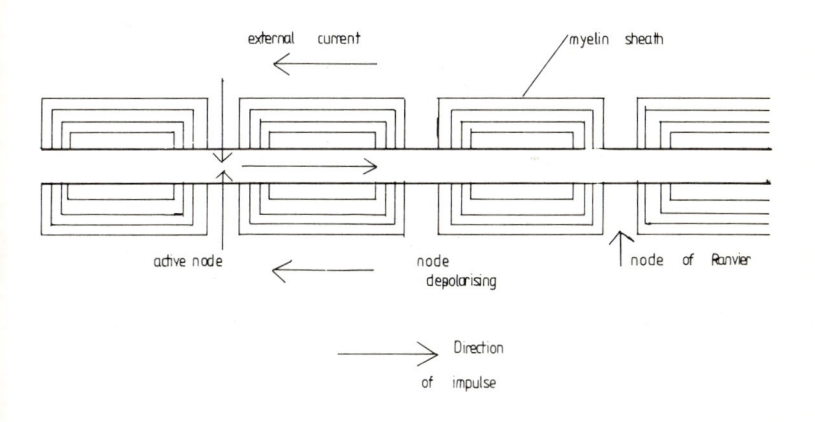

Evidence for saltatory conduction is obtained by measuring external longitudinal currents using two closely-spaced electrodes. Passage of an impulse is registered by a biphasic change in the potential difference between these electrodes. As long as they are both between the same pair of nodes, the time of arrival of the signal does not depend on their position. Only when a node is passed does an extra delay occur. This discontinuous pattern of external current is consistent with the proposed local circuits in myelinated nerve. Local anaesthetics which inhibit impulse propagation by blocking the sodium channel are more effective if applied to the node than when applied to an internodal region. Similarly, local cooling is most effective in causing nerve block if applied at a node.

Higher animals possess the ability to myelinate nerves while many primitive creatures do not. The evolution of myelination may have been essential for the development of large organisms with rapid adaptive reactions. For example if the nerve supply to the human arm consisted of unmyelinated fibres, then to perform the same function, they would be of such bulk as to fill the whole limb with nervous tissue.

4.7 Summary

The properties of the axon membrane which give rise to the phenomenon of nerve impulse propagation may be summarised as follows.

(1) The resting membrane has a low ionic permeability so that electrochemical gradients of ions across it are not immediately dissipated.
(2) The membrane contains a sodium pump which uses the energy of ATP to create and maintain ion concentration gradients. Potassium ions are loaded into the axon while sodium ions are pumped out.
(3) The resting membrane is selectively permeable to potassium ions so that a resting potential of -70 mV is set up which is close to the Nernst equilibrium potential for potassium.
(4) The membrane contains two types of gated ion channel which are opened and closed by changes in membrane voltage.

The sodium channel is opened rapidly by depolarisation and it then inactivates slowly. It is responsible for the fast rising phase of the action potential. The potassium channel is opened slowly by depolarisation and is responsible for the slower falling or repolarising phase of the action potential.

A common mistake made by students of nerve physiology is to confuse the sodium pump and sodium channel. They are quite independent proteins in the axon membrane and Table 4.1 is a summary of their contrasting properties. A squid nerve will conduct many thousands of impulses after the sodium pump has been blocked with ouabain. This is because the sodium pump only creates the necessary ion gradients and a single action potential does not significantly deplete these gradients. Blockage of the sodium channels with tetrodotoxin (TTX), however, causes an immediate and complete loss of nerve function.

Table 4.1: Comparison of Sodium Pump and Sodium Channel (after Keynes, 1975)

	Sodium Pump	Sodium channel
Direction of ion movements	Up electrochemical gradient	Down electrochemical gradient
Source of energy	ATP	Pre-existing electro-chemical gradients
Blocking agents	Ouabain, metabolic inhibitors	Tetrodotoxin Metabolic inhibitors have no effect
Number in squid membrane	4000 per μm^2	500 per μm^2
Rate of ion flux per protein unit	10^2 ions per second low flux	10^5 ions per second very high flux
Effect of membrane potential	None	Opens channel gates

The nervous impulse is the best understood example of cell excitation. It is difficult to overstate the importance of the pioneering work performed by Hodgkin and Huxley on the squid giant axon in the late 1940s and early 1950s. Not only has their model proved applicable in general terms to myelinated mammalian nerves, but its concepts are widely used as the starting point for the investigation of other excitable processes. For example our present models of excitation in skeletal muscle (Chapter 8) and cardiac muscle (Chapter 9) are derived from ideas first introduced by Hodgkin and Huxley. The action potential in nerve is used to convey information across large distances within the organism. It does not convey information between cells but merely from one part of a cell to another part. Use of information in a multicellular organism clearly also requires rapid communication between cells and this is the subject of the next chapter.

Further Reading

Armstrong, C.M. (1974), Ionic pores, gates and gating currents. *Q. Rev. Biophys. 7*, 2.

Cole, K.S. and Curtis, H.J. (1939), Electrical impedance of the squid giant axon during activity. *J. Gen. Physiol. 22*, 671.

Hille, B. (1970), Ionic channels in nerve membranes. *Prog. Biophys. Molec. Biol. 21*, 1.

Hodgkin, A.L. (1964), *The Conduction of the Nervous Impulse.* Liverpool University Press, Liverpool.

— (1964), The ionic basis of nervous conduction. *Science 145*, 1148.

Hodgkin, A.L. and Huxley, A.F. (1952), A quantitative description of membrane current, and its application to conduction and excitation in nerve. *J. Physiol. 117*, 500.

Huxley, A.F. (1964), The quantitative analysis of excitation and conduction in nerve. *Science 145*, 1154.

Katz, B. (1966), *Nerve, Muscle and Synapse.* McGraw Hill, New York.

Keynes, R.D. (1975), Organisation of the ionic channels in nerve membranes. In: Tower, D.B. (ed.), *The Nervous System*, Vol. 1. Raven Press, New York.

— (1979), Ion channels in the nerve-cell membrane. *Sci. Am. 240*, 98.

Noble, D. (1966), Applications of Hodgkin-Huxley equations to excitable tissues. *Physiol. Rev. 46*, 1.

Stevens, C.F. (1978), The neurone, *Sci. Am. 241*, 49.

5 THE SYNAPSE

5.1 Introduction

A synapse is a region where the membranes of two excitable cells are closely apposed and are specialised to allow a transfer of information from one cell to the other. This transfer is usually achieved by a process of chemical transmission. A chemical transmitter substance is released in controlled amounts by one cell and diffuses rapidly across to bind to the membrane of the second cell. The binding of transmitter causes changes in the ionic conductance of the post synaptic cell membrane and thereby affects the fluxes of ions into and out of this cell. By this means activity in the first cell can be sensed by the second cell. Such chemical synapses are unidirectional. One membrane is specialised to release the transmitter substance while the other is specialised to respond to it, so that information is transferred in one direction only.

The purpose of this chapter is to describe the local membrane specialisation at synapses which gives rise to the phenomenon of chemical transmission. Experimentally two preparations have proved most useful in elucidating synaptic mechanisms. These are the synapse between nerve terminals and voluntary muscle in the frog, and the synapse at the origin of the giant axon in the squid. The frog neuromuscular junction is useful as microelectrodes can be inserted into the muscle fibre to measure membrane potential, while in the squid preparation electrodes can be inserted into the transmitter-releasing nerve terminal. Accordingly the description which follows has largely been drawn from these two examples. None the less there is considerable evidence that they represent a general phenomenon. Chemical transmission is a powerful mechanism for integrating information and is the process which appears to underlie many aspects of brain function. The final sections of this chapter are concerned with the potential usefulness of the chemical synapse in complex information processing.

5.2 Synaptic Structure and Components

The functional synaptic structures are so small that they cannot be resolved by light microscopy. Even electron microscopy can only give

58 *The Synapse*

an outline of the synapse, as individual protein and lipid molecules are too small to be identified. A very schematic synapse as shown by the electron microscope is illustrated in Figure 5.1. The membrane from which transmitter is released is known as the presynaptic membrane, while the membrane about 20 nanometres (nm) away across the synaptic gap is the postsynaptic membrane. Close to the presynaptic membrane in the presynaptic cell are numerous synaptic vesicles each of which is enclosed by its own membrane. The chemical transmitter is believed to be stored in these vesicles ready for release into the synaptic gap. There are many different transmitters used by different types of cell. Some chemical transmitters can be shown to reside in the vesicles because they bind fluorescent histochemical markers which are observed to collect in the vesicles. In other cases the evidence is indirect, such as the observation that the transmitter is protected from degrading enzymes unless the vesicles are disrupted by osmotic shock.

Figure 5.1: Schematic Diagram of a Synapse. The lowest picture of exocytosis is consistent with the 'omega figure' seen in electron micrographs.

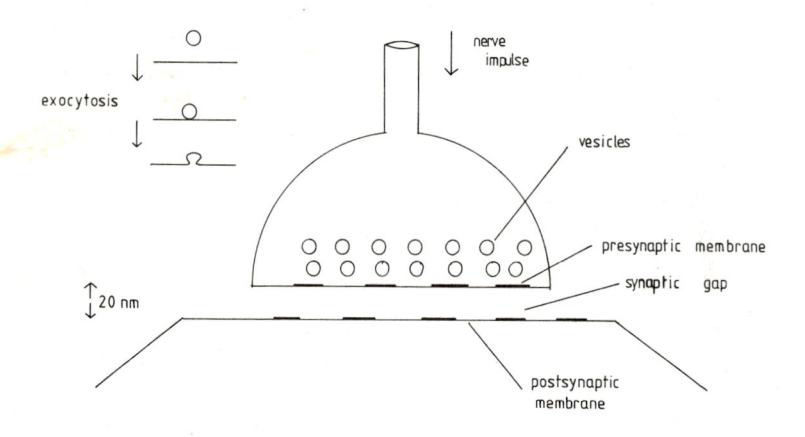

Perhaps the most convincing evidence for the vesicular storage of transmitter comes from experiments in which homogenised fractions of the nerve terminal are separated by ultracentrifugation (see for example von Euler, 1972). One fraction is found to contain vesicles (as visualised by electron microscopy) and this fraction can also be shown to contain most of the transmitter. Such experiments are open to criticism, however, on the grounds that the vesicles seen after fractionation may not be identical to those present *in vivo*. In frog muscle there is strong evidence from postsynaptic measurements (to be discussed in Section 5.4) that transmitter is released in packets. This supports the idea that the frog neurotransmitter is stored in presynaptic vesicles awaiting release. Nevertheless, although the hypothesis of vesicular storage has conceptual appeal, it is not yet well-supported as a ubiquitous phenomenon.

At first glance the storage of transmitter in vesicles in the presynaptic cell is rather odd as it must cross two membranes to get from there into the synaptic gap from which it can influence the postsynaptic cell. The explanation is that transmitter is released by a process of *exocytosis* as illustrated on the left in figure 5.1. The vesicle membrane fuses with the presynaptic membrane and the contents of the vesicle are thereby expelled into the synaptic gap. Sometimes an electron micrograph shows an 'omega figure' appearance which is held to represent exocytosis taking place. The hypothesis of excytosis is difficult to prove but again it is consistent with the observed release of transmitter in packets.

Also present at the synaptic region are enzymes which break down the transmitter after release and so help to reduce the time for which it can act on the postsynaptic membrane. At the frog neuromuscular junction the transmitter is acetylcholine (ACh) and it is destroyed by the enzyme acetylcholinesterase (AChE). This destruction produces the substance choline which is pumped back into the presynaptic terminal for re-use. Replacement of used vesicles is accomplished by two methods. Some are made in the cell body and transported down the axon to the terminal while others are synthesised at the terminal from pinched-off cell membrane.

The pre- and postsynaptic cell membranes also contain specialised excitable ion-channels which are vital to synaptic function. They are not seen in electron micrographs and their properties can only be derived from measurements of ion flow across the synaptic membranes.

5.3 The Control of Transmitter Release

The usual stimulus to transmitter release is the arrival of an action potential at the nerve terminal following propagation along the nerve axon. More precisely it is the depolarisation of the terminal membrane associated with the action potential which triggers release. This membrane depolarisation gates, or opens, voltage-sensitive calcium ion

60 *The Synapse*

channels in the terminal membrane which results in a flux of calcium ions into the presynaptic cell. These channels are illustrated in Figure 5.2, which also shows the resting calcium ion concentrations inside and outside the cell. There is a very large electrochemical gradient favouring calcium ion entry but at rest the membrane is almost impermeable to calcium. The calcium channels opened by a depolarisation of membrane voltage therefore conduct a large current of calcium ions into the cell. Because the resting intracellular concentration of calcium ions is so low, this influx causes a significant change in the internal chemical potential of calcium which in turn causes a massive stimulation of exocytosis, so that approximately 100 vesicles of transmitter are released almost simultaneously.

Figure 5.2: The Role of Calcium Ion Movements Across the Presynaptic Membrane

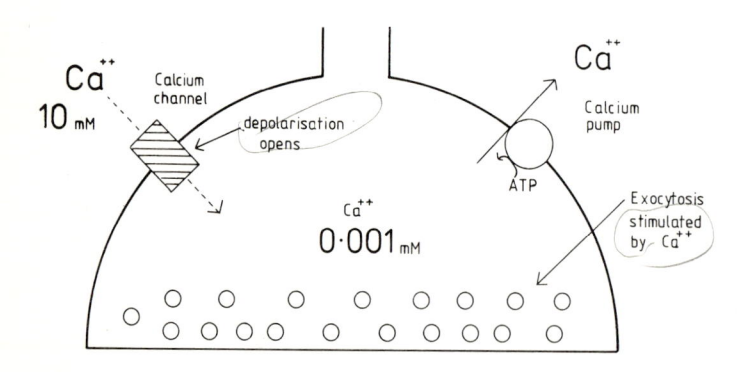

Some evidence for this scheme of excitation-secretion coupling can be obtained by showing that the removal of external calcium ions blocks synaptic transmission. More direct evidence has been obtained in the squid giant synapse by Katz, Miledi and co-workers using a micropipette electrode in the presynaptic cell. The effect of presynaptic depolarisation on transmitter release is monitored by measuring the postsynaptic voltage as shown in Figure 5.3. Terminal depolarisation by current injection in the absence of an action potential will cause release of transmitter as long as external calcium is present. However if the depolarisation is very large and membrane potential reaches +130 mV then no transmitter release is stimulated. This is consistent with the calcium entry hypothesis as 130 mV is close to V_{Ca} (the equilibrium potential for calcium ions across the nerve membrane). No net calcium ion entry would be expected at this potential, even though channels were open. The channels do indeed appear to be open, as when the presynaptic membrane potential returns suddenly from 130 to -70 mV the electrochemical gradient for calcium is restored, and a release of transmitter is seen (as illustrated on the right-hand side of Figure 5.3). More direct evidence for the role of calcium ion entry has been obtained by Miledi (1973) who showed that direct micropipette injection of calcium ions into the squid terminal can cause transmitter release.

Figure 5.3: Transmitter Release at the Squid Giant Synapse Stimulated by Presynaptic Depolarisation and Detached by Changes in the Postsynaptic Voltage

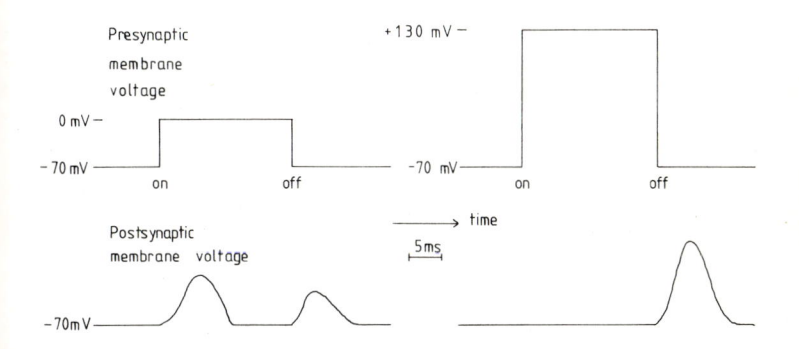

Voltage-sensitive calcium channels are present throughout the nerve cell membrane but are particularly concentrated at the terminal regions. They are not related to the sodium channel of the nerve impulse as tetrodotoxin (TTX) has no effect on them. If depolarisation is maintained they do show a slow inactivation, and they are blocked by magnesium ions. The rise in the terminal free calcium ion concentration is transient as many protein sites in the cell bind calcium with high affinity and rapidly mop-up the extra ions. In the long term, to maintain a steady state, calcium ions must be pumped out of the cell to maintain the electrochemical gradient. This is accomplished by an ATP-dependent pump and may also be associated with a coupled sodium ion movement into the cell.

The mechanism of the coupling between intracellular calcium ion concentration and exocytosis has proved difficult to investigate. In the resting state each vesicle appears to have a small constant probability of release which results in a small resting rate of single vesicles undergoing exocytosis, on average one about every three seconds. When the terminal calcium ion concentration is raised this probability is massively increased so that within one millisecond over 100 vesicles are released. Some results indicate that the probability of release is proportional to the fourth power of calcium concentration, giving rise to suggestions that four calcium ions are somehow required to release one vesicle. Much of the evidence on transmitter release has come from studies of transmitter action on the postsynaptic membrane, and this is the subject to which we now turn.

5.4 The Effect of Transmitter on the Postsynaptic Membrane

The postsynaptic membrane contains ion channels which are controlled by the binding of transmitter substance. These channels are therefore gated by specific chemical stimuli and the ion flows which result enable the postsynaptic cell to respond to the nearby release of transmitter. A significant amplification is achieved by these channels. For example, at the frog neuromuscular junction a single acetylcholine transmitter molecule can open an ion channel which will allow over 100,000 sodium ions to enter the postsynaptic cell.

Figure 5.4: Measurement of Membrane Voltage in a Frog Muscle Cell. Miniature endplate potentials (mepps) are only detected near the endplate.

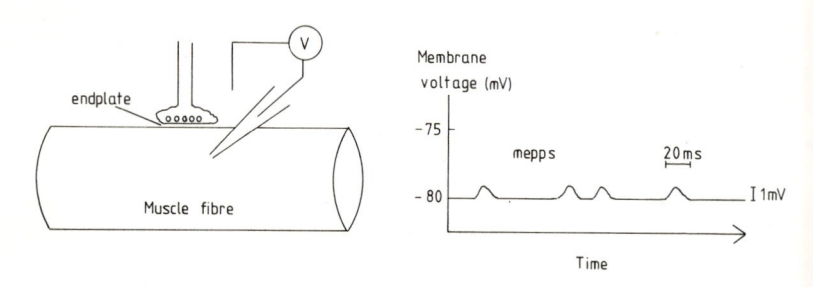

Measurement of local postsynaptic membrane potential changes can be accomplished in frog muscle fibres using micro-electrodes as shown

in Figure 5.4. In the absence of nerve stimulation this potential shows small depolarisations, known as miniature endplate potentials (mepps) as seen on the right of this figure. The trace in Figure 5.5A is the potential change evoked in the muscle membrane by a single stimulation of the motor nerve. The initial response is the endplate potential (epp) which triggers an action potential in the muscle membrane. The mepps and epps are the result of ion flow through channels opened by acetylcholine. A single miniature endplate potential is due to the release of one transmitter vesicle while the endplate potential is due to the simultaneous release of one hundred vesicles evoked by the presynaptic action potential. This common identity can be clearly shown if the nerve is stimulated in the presence of low external calcium ion concentration and high magnesium ion concentration. This reduces presynaptic calcium entry so that only a handful of vesicles are released by the impulse. Recordings of these small endplate potentials show that they are integer multiples of miniature endplate potentials as in Figure 5.5B.

Figure 5.5: A — Frog Muscle Membrane Voltage Measured near the Endplate Responding to Presynaptic Stimulation. The transient signal at 1.5 ms is the stimulus artefact. B — The Quantal Nature of Small Endplate Potentials.

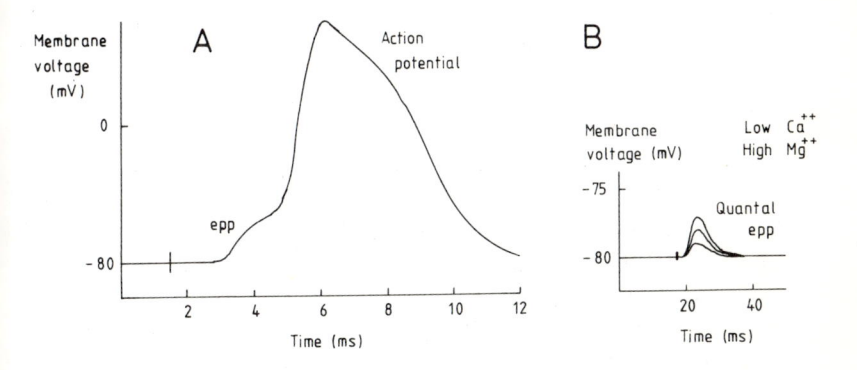

The acetylcholine-sensitive channel of the muscle membrane can be investigated by a crude form of voltage clamp experiment. The channel has a *reversal potential* of about −15 mV. In other words when the membrane potential is below this level the channel conducts a current into the cell, but when membrane potential exceeds it the channel allows an outward current. This is consistent with the idea that the channel is permeable both to sodium and potassium ions. Confirmation is obtained by finding that the reversal potential is a function of both external sodium and potassium ion concentrations, as predicted theoretically by Equation 3.9 (page 34). Recently, further studies have allowed measurement of the single unit conductance of the acetylcholine-sensitive channel. One method involves a statistical analysis of the membrane current 'noise' caused by random opening and closing of channels. Another, pioneered by Neher and Sakmann (1976), uses very sensitive external electrodes to measure the small currents due to single channel openings. The unit conductance is in the range 20-50 picosiemens (pS) which indicates a very high ion flow rate. This channel is therefore likely to be a pore structure, like gramicidin and the ion channels of the nerve membrane. After opening due to acetylcholine binding the channels close again after a period of about one millisecond.

The effect of transmitter release at the neuromuscular junction may be considered as follows. One impulse causes simultaneous release of about 100 vesicles each of which contains some 50,000 acetylcholine molecules. Of the five million transmitter molecules released only 100,000 succeed in opening a postsynaptic channel, but this is sufficient to cause 10,000 million sodium ions to enter the muscle within one millisecond. This inward current produces the endplate potential which exceeds the threshold for initiation of a muscle action potential.

5.5 The Pharmacology of Synapses

A number of substances are known which have selective inhibitory effects on the synaptic processes discussed above. These are important for two reasons: they are of great help in pure research into synaptic mechanisms, and some are important drugs which derive their therapeutic value from specific effects on certain synapses. Using these drugs as experimental tools it is possible to construct a set of criteria which must be satisfied to prove that a particular transmitter is employed at a given synapse.

(1) The presynaptic cell has the ability to synthesise transmitter in sufficient quantities. If synthesis is blocked by a drug then so eventually is transmission through the synapse.

(2) The transmitter is released in markedly larger amounts by the

presynaptic cell during stimulation than in the absence of stimulation.
(3) External application of the transmitter substance produces an appropriate postsynaptic response.
(4) Substances which block transmitter release inhibit transmission through the synapse but do not affect the response to externally applied (exogenous) transmitter.
(5) Substances which inhibit the postsynaptic action of transmitter block both the synapse and the action of exogenous transmitter.
(6) Enzymes or pumps are present which remove the transmitter from the synaptic gap. Blocking these tends to increase the duration of transmitter action.

There is a great variety of different synapses which may connect nerve cell and nerve cell or nerve and muscle or sensory receptor and nerve, and so on. These employ several types of transmitter and many different postsynaptic channels. In only a few cases can a transmitter be proved by these criteria to be present. The best-understood synapse is the frog neuromuscular junction. At other synapses, particularly in the central nervous system, often only one or two of the criteria are met. A large number of substances such as acetylcholine, noradrenaline, dopamine, glycine, glutamate and serotonin (5-hydroxytryptamine) are believed to be central neurotransmitters.

There are nuermous drugs of clinical importance which act on synapses. Drugs such as tranquillisers, antidepressants and sleeping pills, which affect the brain, act at least in part by influencing central synapses. Much more is known about succinylcholine and curare which inhibit synaptic conduction between nerve and voluntary muscle, causing paralysis. They are useful as muscle relaxants during surgery. Propranolol is a drug which inhibits the synapses by which nerves stimulate heart muscle and it therefore reduces cardiac oxygen demand. It is useful therapeutically in reducing blood pressure and in preventing myocardial ischaemia. The mechanism of its therapeutic action appears to involve both cardiac and extra-cardiac sites. For a full account of the many interactions between drugs and synapses the reader is referred to the work of Kruk and Pycock (1979).

5.6 Synaptic Information Processing

The synapses discussed so far have all been excitatory; that is, activity in the presynaptic cell tends to excite activity in the postsynaptic cell. There exists a quite separate class of *inhibitory* synapses where presynaptic activity tends to reduce postsynaptic activity. The difference resides in the ion selectivity of the channels in the post synaptic membrane opened by the transmitter. Consider the inhibitory synapse in Figure 5.6 in which the transmitter opens potassium-conducting channels in the postsynaptic membrane. Here the membrane potential will become more negative as these channels open and potassium ions flow out of the cell.

Figure 5.6: Hypothetical Nerve Cell which Integrates One Excitatory and One Inhibitory Input. Stimulation of the inhibitory input produces an inhibitory postsynaptic potential (ipsp).

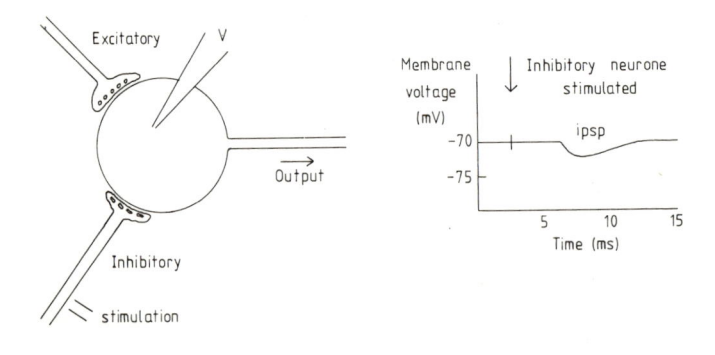

Generally speaking, an excitatory synapse is one at which current is induced to flow into the cell causing depolarisation, while an inhibitory synapse is one where current is induced to flow out causing hyperpolarisation. This transient transmitter-induced hyperpolarisation is known as the inhibitory postsynaptic potential (ipsp) and is shown on the right of Figure 5.6. The direction of current flow is determined by the reversal potential of the ion channels opened by the transmitter. At an excitatory synapse the reversal potential is greater than the threshold membrane voltage for excitation of an action potential, while at an inhibitory synapse it is below that threshold. In mammalian nerve cells channels primarily permeable to potassium or chloride ions are inhibitory, while those primarily permeable to sodium are excitatory. This follows from the fact that V_K and V_{Cl} are below the threshold membrane voltage while V_{Na} is above it.

We can now see, in theory, how a cell can integrate information from two or more sources. Figure 5.6 shows a nerve cell which receives two synaptic inputs, one excitatory and one inhibitory. Whether this cell is excited or not is controlled by whether the inward ionic current caused by the excitatory transmitter exceeds the outward current caused by the inhibitory transmitter. The rate at which the cell fires depends on the amount by which current influx exceeds efflux and this in turn depends on the impulse frequencies of the two inputs. The example is purely theoretical as real nerve cells are much more complex. In the first place a single nerve cell can receive thousands of inputs, and secondly these inputs are not all symmetrically distributed on the cell body, but synapse onto the fine branching processes or dendrites which are attached to the cell. The complex cell geometry means that the relationship between inputs and output may be extremely complicated.

Some simple examples of information processing by nerve networks can be explained on the present restricted model. In the cat spinal cord stimulation of sensory nerves will give rise to reflex muscular responses via motor nerves. *Spatial summation* occurs when stimulating two sensory nerves together produces a response which neither alone can elicit. This may be due to these nerves both having an excitatory synapse onto the same motor nerve cell. *Temporal summation* occurs when stimulating the same nerve twice in rapid succession produces a response where a single stimulus elicits none. This is associated with the maintained postsynaptic depolarisation caused by one impulse, to which the next impulse may add and thereby cause membrane voltage to exceed the threshold for impulse generation.

Not all synapses are chemical in nature; there are also electrical synapses which operate by allowing some direct electrical continuity between the two cells involved. These synapses are not unidirectional and some are present in, for example, the retina. Their importance in the rest of the central nervous system is unknown. Another variation is present in synapses where transmitter release is controlled not by presynaptic action potentials but by graded depolarisations. The potential complexity of the synaptic mechanism is very great, and it is possible to argue that even the highest aspects of brain function can be ascribed to synaptic processes.

5.7 Summary

The sequence of events in chemical synaptic transmission is illustrated in Figure 5.7 and may be briefly stated as follows.

(1) The presynaptic terminal is invaded by an action potential and the associated depolarisation opens calcium ion channels in the terminal membrane.

(2) Entry of calcium ions down a large electrochemical gradient increases the internal calcium concentration and stimulates a transient exocytosis of transmitter into the synaptic gap.

(3) Transmitter diffuses rapidly to specific protein binding sites (receptors) on the postsynaptic membrane where it opens ion channels. Ionic currents through these channels alter the membrane potential of the postsynaptic cell in a direction determined by the ion selectivity of the channels concerned.

(4) A depolarisation of the postsynaptic cell causes excitation while a hyperpolarisation causes inhibition.

Figure 5.7: Summary of Chemical Synaptic Transmission

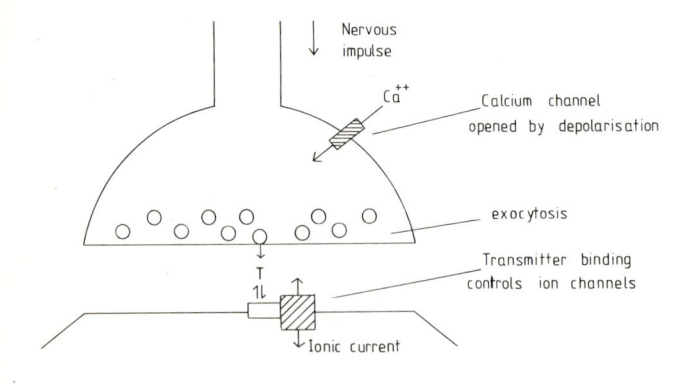

In addition there are high-affinity calcium ion binding sites in the terminal cell which ensure that the rise in internal calcium concentration is transient. Similarly there are enzymes which remove transmitter from the synaptic gap so that the synapse is rapidly reprimed for response to another impulse. There is an immense variety of different types of synapse between different classes of cell, using different transmitters and different postsynaptic channels. Many of these can be distinguished pharmacologically and this has considerable therapeutic importance. Because nerve cells exhibit complex geometry and have multiple inputs the synapse is a powerful information processing mechanism.

At both the presynaptic and postsynaptic membranes signal amplification is achieved by ion channels using energy stored in ionic electrochemical gradients. The voltage-sensitive calcium ion channel utilises the large calcium ion gradient across the presynaptic membrane. Transmitter-gated channels in the postsynaptic membrane use sodium, potassium and chloride ion gradients. There is considerable research interest in the acetylcholine-sensitive channel of muscle membranes, and two complementary techniques are proving effective. First there is the measurement of the functional properties of the channel by electrophysiological measurements and secondly the extraction of acetylcholine-binding protein using radioactive ligands followed by its biochemical characterisation. A future goal is to obtain detailed structural information on the protein which might then be related to its function as a gated ion channel.

This chapter and the previous one have discussed how the organism communicates and integrates information. We now turn to the question of how this information is obtained in the first place by the organs of sensation.

Further Reading

Anderson, C.P. and Stevens, C.F. (1973), Voltage clamp analysis of acetylcholine-produced endplate current fluctuations at the frog neuromuscular junction. *J. Physiol. 235,* 655.
Baker, P.F. (1972), Transport and metabolism of calcium ions in nerve. *Prog. Biophys. Molec. Biol. 24,* 177.
Euler, V.S. von (1972), Adrenergic nerve particles in relation to uptake and release of neurotransmitter. *J. Endocrinol. 55,* ii.
Hubbard, J.I. (1973), Microphysiology of vertebrate neuromuscular transmission. *Physiol. Rev. 53,* 674.
Katz, B. (1966), *Nerve, Muscle and Synapse.* McGraw Hill, New York. (Chs 8-10.)
Katz, B. and Miledi, R. (1967), A study of synaptic transmission in the absence of nerve impulses (squid). *J. Physiol. 192,* 407.

Kruk, Z.L. and Pycock, C.J. (1979), *Neurotransmitters and Drugs*. Croom Helm, London.

Miledi, R. (1973), Transmitter release induced by the injection of calcium ions into nerve terminals. *Proc. Roy. Soc. B 183*, 421.

Neher, E. and Sakmann, B. (1976), Single-channel currents recorded from membrane of denervated frog muscle fibres. *Nature 260*, 799.

Rang, H.P. (1975), Acetylcholine receptors. *Q. Rev. Biophys. 7*, 283.

Stevens, C.F. (1978), The neurone. *Sci. Am. 241*, 49.

Part Three

THE ACQUISITION OF INFORMATION

6 VISION

6.1 Introduction

The eye is man's most sophisticated sensory organ. It possesses an optical system of variable focusing power which creates an image of the visual world on the retina lying at the back of the orbital cavity. The retina is a thin layer of excitable tissue about five cells deep which performs the vital process of sensory transduction. In other words it converts the visual image into patterns of electrical activity within its excitable cells, and thereby enables the information contained in that image to be assimilated by the nervous system. The visual information is then passed to higher centres of the brain along the million individual axons which make up each optic nerve.

There are a number of different cell types in the retina but the two which will be discussed here are the *rods* and *cones*. These are the photoreceptor cells and only they have the ability to convert light energy into electrical activity. The other cells of the retina are important in amplifying the electrical signals and in integrating information by synaptic mechanisms to enhance contrast and detect specific patterns. The object of this chapter is to give an account of the process of transduction from light into electrical activity. This is the stage at which information about the external world is acquired by the organism. Phototransduction is a good example of the type of specialisation which allows an excitable cell to be exquisitely sensitive to a single modality of stimulus, in this case to light.

Light produces dissimilar responses in vertebrate and invertebrate photoreceptors, and only the former are considered here. Experiments on vertebrates such as the mudpuppy, turtle and rat show considerable agreement in their results and the information which follows has been obtained using these animals. An understanding of the properties of photoreceptors (and their distribution in the retina) leads to explanations for many visual phenomena such as colour vision, high acuity vision and vision at very low light levels.

6.2 The Structure of the Retina

Figure 6.1: Schematic Diagram of the Cellular Structure of the Retina

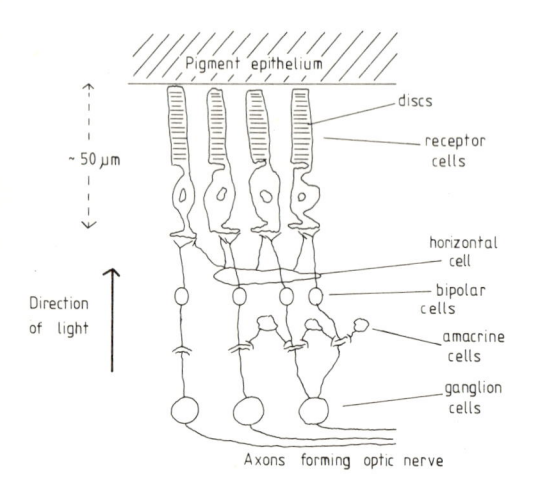

The cellular structure of the retina is shown schematically in Figure 6.1. Note how the light must pass through the outer cell layers before reaching the receptors where it is detected and produces characteristic electrical changes. The electrical activity of the photoreceptors is transferred via synaptic connections to the bipolar and ganglion cells, and these latter cells have axons which make up the optic nerve. Horizontal and amacrine cells are responsible for lateral synaptic interactions and each of these cells makes synaptic contacts across a considerable area of retinal surface. The two types of photoreceptor, rods and cones, are not distributed evenly across the retina. Cones are present only in a small area on the optical axis known as the fovea, while rods cover the rest of the retinal surface. Cones are responsible for high acuity colour vision while rods are active at low levels of illumination. In the foveal region there is a high density mosaic of cones with a separation of about $2\mu m$ between receptors. This distance is equivalent to a separation in visual space of 0.5 minutes of arc and is one of the factors which limit the resolving power of the eye to objects greater than one minute of arc apart. The small area of the fovea means that high acuity colour vision is only possible in the centre of the visual field but this fact is not subjectively obvious because of the rapid scanning motions of the eye.

The structures of rod and cone receptors are illustrated in Figure 6.2 and, apart from the different shapes, the major distinction is in the stacked membrane discs which fill the outer segment of each cell. The rod discs are quite separate from the cell membrane and enclose membrane-bound spaces inside the cell, while in cones the discs are formed by invaginations of the cell membrane. Both rods and cones contain pigment molecules which are attached to the discs as intrinsic membrane proteins and which are responsible for the first step in phototransduction.

Figure 6.2: Comparison of Rod and Cone Structure

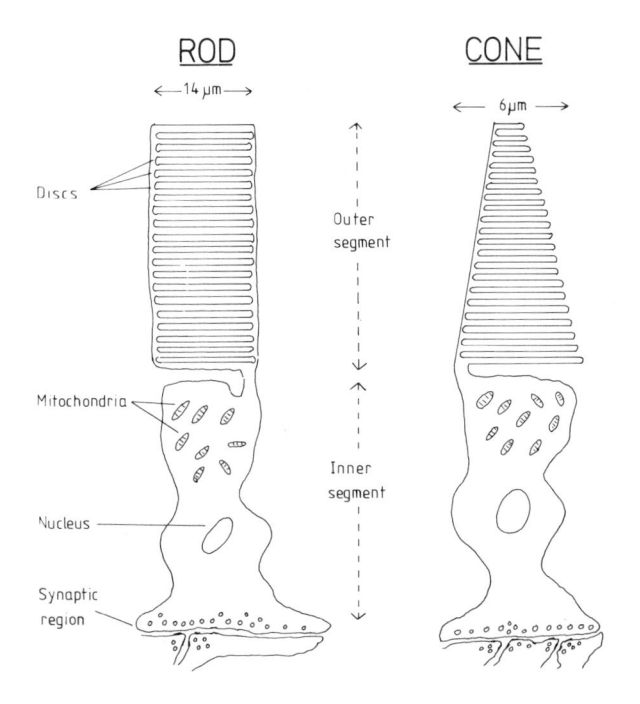

6.3 The Photopigment Rhodopsin

The rod pigment is a purple protein called rhodopsin which can be shown to be associated with the disc membranes by labelling it with specific fluorescent antibodies. It consists of a protein, opsin, to which is covalently attached the aldehyde of vitamin A (retinal) as shown in Figure 6.3. The absorption spectrum of rhodopsin is shown in Figure 6.4 and it exhibits a peak at 500 nm, indicating that light of this wavelength is most effectively absorbed by the pigment. Rhodopsin is highly photolabile and is rapidly bleached by light, the action spectrum for which also has a peak at 500 nm. Another action spectrum can be obtained by comparing the effectiveness of light of different wavelengths in producing optic nerve activity. Again a single peak is found at 500 nm. Similarly the spectral sensitivity of the human eye at low light levels has the same peak. Finally Rushton (1972) has demonstrated that the action spectrum for bleaching of human pigment *in vivo* is also similar to Figure 6.4. This coincidence of spectra is the best evidence that light absorption by rhodospin underlies photoreception by the rods.

Figure 6.3: The Isomerisation of 11-cis Rhodopsin by Light. The subsequent reactions are not photochemical

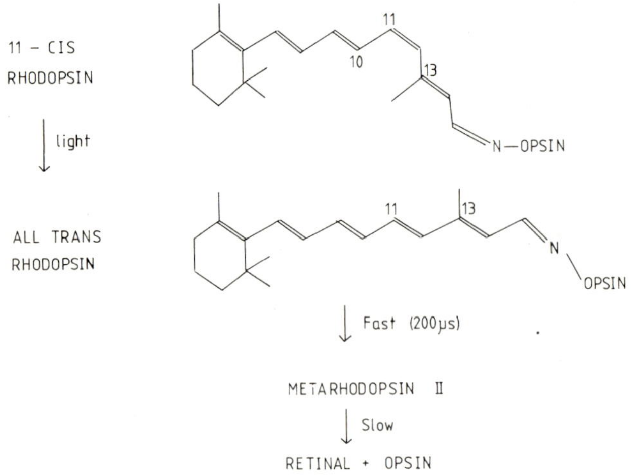

In the dark the retinal in rhodopsin is in the 11-*cis* conformation and the first event caused by light absorption is an isomerisation to the *all trans* form as in Figure 6.3. The 11-*cis* isomer is inherently unstable as steric interactions between the methyl group on carbon 13 and the hydrogen atom on carbon 10 necessitate a twisting of the molecule.

This twist is relieved as the molecule flips out to the straight *all trans* form. Therefore energy is released by the absorption of light, and this may be the first step in signal amplification. Following isomerisation the rhodopsin molecule undergoes a sequence of very rapid chemical reactions to form an intermediate called metarhodopsin II. This occurs within 200 μs but further reactions proceed only slowly, with a time course of minutes, and eventually retinal and opsin dissociate. Enzymes are present which slowly resynthesise the 11-*cis* form of rhodopsin. After reception of light there is a latent period of 100 milliseconds before significant electrical changes take place in the photoreceptor cell. Therefore although rhodopsin absorption of light leads to vision, none of the known chemical reactions of rhodopsin can explain the precise coupling of absorption to electrical events.

Figure 6.4: The Absorption Spectrum of Rhodopsin Compared with Human Visual Sensitivity

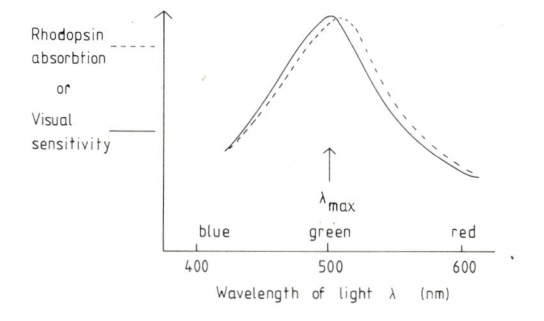

The high concentration of rhodopsin in disc membranes and the stacking of discs in the rod outer segment make this a good preparation for structural analysis by X-ray diffraction. Rhodopsin behaves approximately as a sphere with a hard core diameter of 4.5 nm which is mobile in the plane of the disc membrane, as if it were in a two-dimensional liquid. It has nearest-neighbour molecules at about 6.0 nm. The absorption of light by rhodopsin *in situ* on the disc membranes is maximal if the light is directed axially into the rod and passes perpendicularly through the discs. Light directed across in the plane of the disc is only absorbed one-sixth as efficiently. This means that the rhodopsin is unable to rotate about any axis in the plane of the membrane (in other words it cannot tilt its axis perpendicular to the membrane) as would be predicted for an intrinsic membrane protein. Clearly this orientation facilitates the absorption of light arriving at the receptor from the lens of the eye and inhibits the absorption of stray reflected light. The disc membranes therefore perform an important role in orientating the photopigment.

A rhodopsin-like protein known as bacteriorhodopsin is present at high concentration in the purple membrane of the bacterium *Halobium halobius*. It uses light energy to transport protons across the cell membrane and (as mentioned in Chapter 2) its structure has been estimated at 0.7 nm resolution by Henderson and Unwin (1975) using low-angle electron microscopy. It consists of seven helical sections which span the lipid bilayer membrane and the pigment molecules are arranged in groups of three. By analogy with this molecule it has been suggested that rhodopsin itself might span the disc membrane and form an ion channel which is opened or closed by light. There is no direct evidence for this hypothesis but it might help to explain the coupling of light absorption and electrical changes in the receptor cell.

6.4 Electrical Events Following Photoreception

Some electrical effects can be measured using external electrodes to record the gross electrical potential difference across the retina. A very small signal known as the early receptor potential (ERP) is produced by light stimulation with no delay. Its time course is independent of the wavelength of light used and its amplitude is proportional to the number of photoisomerisations taking place. Bleaching of pigment reduces the ERP which is considered to be due to charge reorientations taking place in the disc membranes as isomerisation changes the conformation of rhodopsin. However the charge flow associated with the ERP is very small indeed (at about one electron per 100 isomerisations) compared to the subsequent charge movements across the photoreceptor cell membrane.

The development of very fine microelectrodes tapered to less than 10 μm at their tips has enabled intracellular electrical recordings to be made from vertebrate photoreceptors. Light causes a membrane hyperpolarisation comprising an initial peak and subsequent plateau as shown in Figure 6.5C. This response has a latency of about 100 ms and is associated with a decrease in total membrane conductance. Reducing the extracellular sodium ion concentration produces a parallel loss in magnitude of the light-induced receptor hyperpolarisation, indicating that sodium currents across the receptor membrane may be involved. Another method of reducing the electrical response to light is to depolarise the photoreceptor by injection of current through a microelectrode. At zero membrane voltage the response is abolished. Taken together these phenomena suggest that the hyperpolarising response is due to a decrease in the sodium ion conductance of the receptor membrane, which causes inhibition of an inward sodium current. Depolarisation inhibits the response as the electrochemical force driving sodium influx is reduced.

Figure 6.5: A — The Dark Current in Rat Rods Driven by Membrane Pumps of the Inner Segment. B — The Current Produced by Light. C — The Hyperpolarising Response to Light in a Turtle Cone.

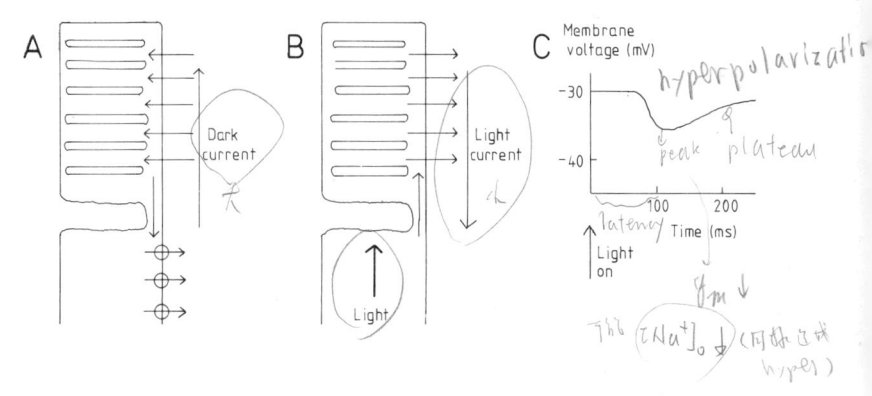

There is a large signal amplification at this stage as in turtle cones a single photoisomerisation produces a voltage change of 30 μV for one-fifth of a second. This represents either the removal of 500,000 positive ions from the cell or the prevention of the same number entering. At high light intensities the hyperpolarising response saturates and its amplitude is not proportional to the number of photoisomerisations. For example, as few as 1,000 isomerisations in a receptor containing 10^8 rhodopsin molecules will produce a half-maximal hyperpolarisation.

The flow of current in photoreceptors has been investigated in rat rods by Hagins and colleagues (1970), using fine external electrodes to detect charge movements. In the dark resting state there is a flow of 'dark current' around the rod as shown in Figure 6.5A. On illumination a 'light current' is produced which travels in the opposite direction (Figure 6.5B). The light current never exceeds the dark current and appears to be due to a local inhibition of the dark current entry into the outer segment in the vicinity of the discs on which light is falling. These results are consistent with the light-induced receptor hyperpolarisation, and confirm the idea that a photoisomerisation in the disc membrane leads to a decrease in the sodium conductance of the nearby cell membrane. The function of the dark current is not clear. It may perform an intracellular transport function by moving charged molecules within the receptor. The current is driven by ion pumping at the inner segment membrane and the electrochemical gradients across the outer segment membrane created by these pumps are one energy source for signal amplification. In one sense the dark current is a direct current signal which reports changes in the conductance of the outer segment membrane to the synaptic region.

Figure 6.6: Hypothetical mechanism for Production of an Electrical Response to Light Absorption

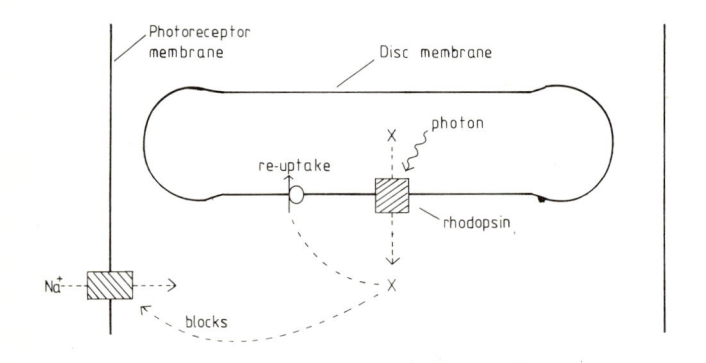

The mechanism by which light absorption by a rhodopsin molecule causes changes in the sodium conductance of the cell membrane has been the subject of considerable recent interest. In rods there is no electrical coupling between disc and cell membranes and the absorbing pigment may lie as far as 5μm from the cell membrane. One explanation is that a chemical transmitter substance is released by the discs and diffuses to the cell membrane where it influences sodium ion channels. This idea appears capable of explaining both the signal amplification at this stage and the delay between light absorption and electrical effects. Consider the model of Figure 6.6. Here light absorption by photopigment opens a channel in the disc membrane allowing ions (X) to flow passively from the disc space into the receptor cytoplasm. These ions then diffuse to the cell membrane and block the entry of sodium ions. Amplification is achieved at two stages. First a single pigment channel can allow many blocking ions to leave the disc in a few milliseconds. Secondly each blocking ion can block a sodium ion channel and prevent the entry of many thousands of sodium ions.

There is some evidence that calcium ions may be the internal transmitter (X). The discs contain enough calcium for this purpose. They have been shown to release calcium as a response to light and sequester it in the dark. Direct injection of calcium into toad rods through a micropipette causes hyperpolarisation. Furthermore light produces an increase in disc membrane conductance as required by the model. Unfortunately these experiments are not easy to perform or interpret, and some researchers have been unable to reproduce the above findings. In cones there is evidence against the calcium hypothesis. The intradisc space and extracellular fluid are continuous in cones, and the hypothesis therefore predicts that external calcium ions should be essential for light-induced electrical responses. This is not found to be the case and pigeon cones will continue to function in media containing no calcium ions and a calcium-chelating agent.

An alternative hypothesis is that cyclic guanosine monophosphate (cGMP) acts as the internal transmitter. High concentrations of cGMP in rods are associated with increased cell membrane conductance. The disc membrane contains an enzyme (phosphodiesterase or PDE) which hydrolyses cGMP and which is activated by light. It is proposed that cGMP acts to open sodium channels in the cell membrane, and that light reduces the cytoplasmic concentration of cGMP by activation of PDE in the disc membrane. The rate at which light induces hydrolysis of cGMP is fast enough to explain both the time course of the electrical response and the signal amplification. At present the competing calcium and cGMP hypotheses remain unproven, but work now in progress should resolve this question in the near future. One important test is to see whether artificial changes in cGMP concentrations can mimic the electrical responses to light.

6.5 Visual Sensitivity

Light rays behave as if they are both waves of electromagnetic radiation and formed from a stream of particles or photons. Light of a given wavelength is associated with photons of a certain energy and the smaller the wavelength the higher the energy. A single photon has a very small amount of energy; for example there are 6×10^{-19} J in a green photon. A photon is known as a quantum of light. At low levels of radiation it is significant that these quanta or packets are indivisible, i.e. no fractions of a photon occur. The human eye is remarkably sensitive and vision is possible in illumination so dim that only 50 photons are being absorbed by the retina each tenth of a second. A reliable sensation of light can be produced by the absorption of only six photons within one-tenth of a second over a retinal area or 20,000 square microns. As this area contains about 5,000 rods it is extremely unlikely that one rod would receive two photons, and it must be concluded that a rod can signal the photoisomerisation of a single rhodopsin molecule. In the turtle retina a single absorption is estimated to cause a hyperpolarisation in rods of 0.7 mV which presumably is sufficient to alter the transmitter release from the synaptic region. There is also a convergence of information from many rods onto a single ganglion cell which helps to increase the reliability of the detection system. The ganglion cell must distinguish random changes in photoreceptor potential from the small burst of signals due to a very weak flash of light.

6.6 Colour Vision

Figure 6.7: Absorption Spectra for the Blue, Green and Red Cone Pigments. The individual cone sensitivities are consistent with these spectra.

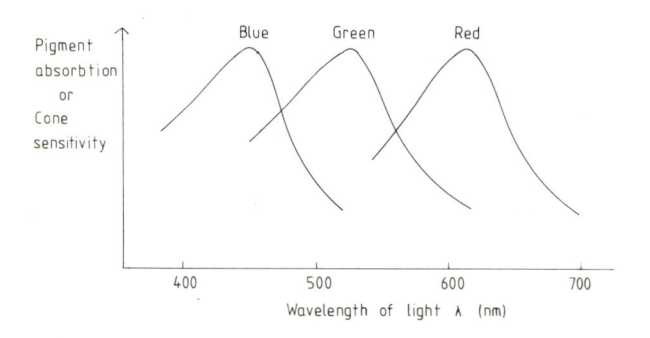

The phenomenon of colour vision results from the presence of three types of cone in the retina, each characterised by a different peak of spectral sensitivity. This trichromatic theory was first proposed by Young (1802) and has received support from recent work on the spectral sensitivities of individual cones. Extracted cone pigments can be divided into three types by their absorption spectra as shown in Figure 6.7. These spectra agree with the absorption properties of single cones. The action spectra of individual turtle cones can be obtained using intracellular recording and again these agree with the graphs in the figure. In man, two out of the three types of cone can be bleached using light of appropriate wavelengths and the spectral sensitivity then measured is due to the third cone type. All cone pigments possess the same retinal molecule as vertebrate rhodopsin and their different spectral peaks are due to a variation in the protein environment around the chromophore. The pigments in the red, green and blue cones are listed in Table 6.1.

Table 6.1: The Cone Pigments

Pigment	λ max (nm)	Colour detected	Absence in man
Erythrolabe	620	red	1%
Chlorolabe	520	green	1%
Cyanolabe	450	blue	rare

In the turtle retina the pigment in the green cones is the same as that employed in the rods yet the time course of the electrical response in these two receptor types is quite different. This further illustrates that the electrical events in the photoreceptor are not directly explained by a chemical reaction of the pigment molecule. The signal produced in a single cone is not dependent on the wavelength of the light absorbed. For example a green cone will respond in precisely the same manner if it captures a blue photon or a green one. This is the principle of univariance. It can be tested by showing that the electrical response of a green cone to a flash of green light can be precisely matched in amplitude and time course using a brighter flash of blue light. A single population of cones does not confer colour vision. Blue light is distinguished from green light by the differential activation of blue and green cones. Therefore the colour image we have of the visual world is only created once information from cones of different types is integrated by synaptic mechanisms. In lower animals this integration can be shown to begin in the retina and even at the level of the cones themselves, but the most important processing, especially in primates, occurs in the visual cortex of the brain.

Colour blindness may result from a defect in one of the cone systems so that the individual only has two fully functioning colour channels, i.e. he is a dichromat rather than a trichromat. Absence of a cone pigment is found in males with the incidences shown in Table 6.1 due to a defective gene on their X chromosome. A further eight per cent of the male population have some other defect in the red or green cone system and they are known as anomalous dichromats. These red-green colour-blind people are clearly disqualified from certain professions and may have considerable trouble interpreting traffic lights and other coloured signals. The condition can be routinely detected using charts containing symbols hidden in a pattern of coloured dots.

6.7 Summary

The detection of light by the photoreceptor cells of the retina has been described. Two types of receptor exist: rods which cover most of the retina and are used in low levels of illumination (scotopic vision), and cones which are concentrated at the fovea and subserve high acuity colour (photopic) vision. These cells contain many hundreds of orientated discs formed from membranes rich in photopigment protein molecules. The rod pigment rhodopsin undergoes a rapid chemical structural reorientation on absorption of a photon which is the first step in light detection. Following this, with a latency of about 100 milliseconds, there is a hyperpolarisation of the receptor cell membrane which affects the rate of transmitter release by the synaptic region of the receptor and is thereby sensed by the neighbouring retinal cells.

The light-induced membrane hyperpolarisation is due to a local decrease in the membrane sodium ion conductance, occuring close to the illuminated discs. A possible explanation for the coupling of light absorption and electrical events is that rhodopsin forms a channel in the disc membrane which allows blocking particles, perhaps calcium ions, to diffuse from the disc space to the surface membrane and there inhibit a sodium entry channel. This model can account for the large amplification in the process by which a single photon prevents half a million sodium ions entering the cell.

Visual sensitivity in the dark is associated with two factors. First a rod is capable of signalling the absorption of a single photon which changes its membrane potential by about 1 mV. Secondly information from hundreds of rods converges through synaptic attachments in the retina onto a single ganglion cell. Colour vision is provided by the red, green and blue cones which contain photopigments tuned to three different wavelengths. Integration of information from these three channels enables a colour image to be formed in the higher centres of the brain.

Further Reading

Barlow, H.B. and Fatt, P. (eds.) (1977), *Vertebrate Photoreception.* Academic Press, London and New York.

Baylor, D.A., Hodgkin, A.L. and Lamb, T.D. (1974), Reconstruction of the electrical responses of turtle cones to flashes and steps of light. *J. Physiol. 242,* 759.

Brindley, G.S. (1970), *Physiology of the Retina and Visual Pathways.* Edward Arnold, London.

Cone, R.A. and Dowling, J.E. (eds.) (1979), *Membrane Transduction Mechanisms.* Raven Press, New York.

Davson, H. (1972), *The Physiology of the Eye.* 3rd edn, Academic Press, New York.

Ernst, W. (1979), The internal transmitter of vertebrate photoreceptors. *Photobiol. Bull. 1,* 128. (cGMP idea.)

Hagins, W.A., Penn, R.D. and Yoshikama, S. (1970), Dark current and photocurrent in retinal rods. *Biophys. J. 10,* 80.

Henderson, R. and Unwin, P.N.T. (1975), Three dimensional model of purple membrane obtained by electron microscopy. *Nature 257,* 28.

Korenbrot, J.I. (1977), Ion transport in membranes: incorporation of biological ion-translocating proteins in model membrane systems. *Ann. Rev. Physiol. 39,* 19.

Liebman, P.A. and Pugh, E.N. (1979), The control of phosphodiesterase in rod disc membranes: kinetics, possible mechanisms and significance for vision. *Vison Res. 19,* 375.

Rushton, W.A.H. (1972), Pigments and signals in colour vision. *J. Physiol. 220,* 1-31P.

Tomita, T. (1970), Electrical activity of vertebrate photoreceptors. *Qu. Rev. Biophys. 3,* 179.

Wald, G. (1968), The molecular basis of visual excitation. *Nature 219,* 800.

Young, T. (1802), On the theory of light and colours. *Phil Trans R. Soc. 1,* 20. Reprinted in *Lectures on Natural Philosophy,* Vol. 2. 1st edn (London, 1807), 613.

7 MECHANORECEPTION

7.1 The Detection of Mechanical Stimuli

Mechanoreceptors are sensory organs which transduce mechanical stimuli into electrical nerve cell activity. They obtain information about both the external environment and the internal organs of the body. For example, the senses of touch and hearing rely on mechanoreception of external stimuli, while information about blood pressure, joint position and muscle length is obtained from internal mechanoreceptors. This type of information is of great value in controlling bodily function. Blood pressure receptors enable appropriate responses to maintain brain perfusion when moving from a lying to standing position; touch receptors on the hand are essential to normal fine manipulation of small objects; and muscle length receptors are required for the control of all movements.

Figure 7.1: Different Degrees of Specialisation in Mechanoreceptors

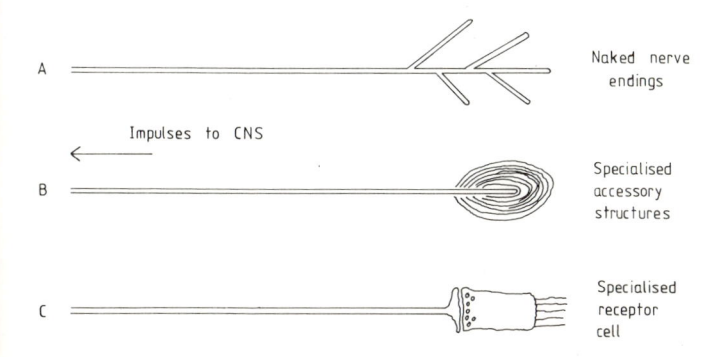

A — Naked nerve endings

Impulses to CNS

B — Specialised accessory structures

C — Specialised receptor cell

There is a large variety of mechanoreceptors which may be broadly classified according to the specialisation of their structures as illustrated in Figure 7.1. The simplest receptors are the naked nerve endings without accessory structures which are found, for instance, in the skin and in the cornea of the eye. More complex receptors are nerve endings surrounded by accessory cells, and the pacinian corpuscle illustrated here (**B**) is found both in the skin and in deeper layers around the internal organs. Finally there are specialised non-nervous receptor cells which perform the transduction from mechanical to electrical activity and synapse onto a nerve terminal. This increasing complexity enables the receptor to be more discriminating about the stimuli to which it will respond. In the cornea only naked nerve terminals are present and they can be stimulated not only by mechanical stimuli but also by thermal and chemical changes. At the other end of the spectrum the inner ear contains hair cells which are only sensitive to the fluid motions caused by transmission of sound from the eardrum. All these receptors transmit their information to the central nervous system (**CNS**) in the frequency and patterns of action potentials travelling along their nerve axons. These afferent nerve fibres often branch extensively so that the activity of many receptors is summed in a single fibre, which is then said to have a *peripheral receptive field* of the area covered by its receptors.

Figure 7.2: Measurement of Electrical Response to Mechanical Stimulation of the Pacinian Corpuscle. The generator potential is revealed by using TTX to inhibit the action potential.

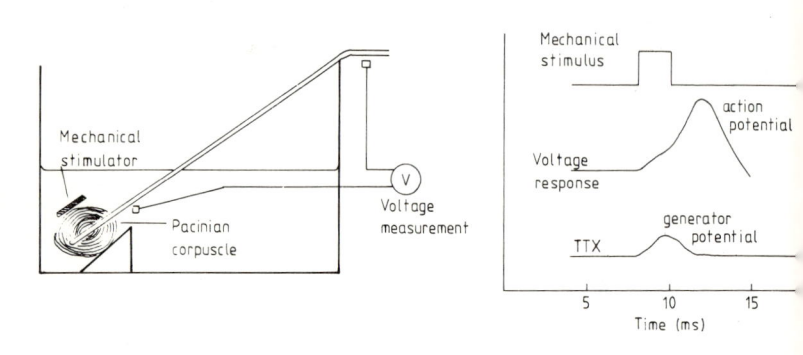

In this chapter two topics will be discussed using selected mechanoreceptors as examples. First the process by which mechanical forces are transduced into electrical changes, and secondly the relationship between the pattern of the mechanical stimulus and the form of the electrical responses of the receptor. This relationship is the input-output function of the receptor, and its most important aspect is the phenomenon of *adaptation* by which a sustained mechanical stimulus only produces a transient electrical response. This is a type of information processing at the level of the receptor by which the brain receives data about transient mechanical events but is not constantly informed of unchanging stimuli. For example, it is not important to be constantly reminded of the pressure of the clothes on our backs. The best understood examples of mechanoreception fall into the intermediate category of complexity of a nerve ending with accessory structures. They are the pacinian corpuscle which is pressure- and vibration-sensitive, and the crayfish muscle stretch receptor. The highly specialised mechanoreceptors, possessing separate receptor cells which are involved in hearing and the sensations of balance, will not be discussed here, and the reader is referred to textbooks of sensory physiology for a full treatment of them.

7.2 The Generator Potential

Mechanical transduction can be considered in three phases: (1) the transfer of the mechanical stimulus through the accessory structures to the nerve terminal itself (2) the production of a graded electrical response (known as the *generator potential* or *receptor potential)* to the mechanical stimulus at the terminal, and (3) the initiation of action potentials in the nerve axon by the generator potential. Strictly speaking it is at phase (2) that the actual process of transduction occurs, and in order to investigate this process it is necessary to remove the accessory structures and selectively inhibit the action-potential mechanism. This aim was substantially achieved by Loewenstein using the pacinian corpuscle preparation illustrated in Figure 7.2. Corpuscles as large as 1 mm in diameter can be dissected from cat mesentery and extracellular electrical recordings made while quartz crystals are used to provide controlled mechanical stimulation. In these experiments short-lived taps of pressure were employed rather than sustained changes in pressure. The electrical responses to these transient stimuli are not altered by removing almost all of the 'onion skin' accessory structures of the corpuscle. The graded electrical responses can be examined in the absence of nerve impulses by inhibiting the action potential using one of several methods. One possibility is to use such small pressure changes that the generator potential never exceeds the threshold for action potential generation; other methods involve blocking the action potential by compressing the first node of Ranvier, by adding a local anaesthetic such as procaine, or best of all by adding tetrodotoxin (TTX) to block the voltage-sensitive sodium channels. The generator potential produced by mechanical stimulation is shown in Figure 7.3 and is a transient depolarisation of the nerve terminal membrane. It does not originate in the accessory structures as they can be removed, and it is not dependent on the ion channels of the nerve impulse as they can be blocked without affecting it. The potential appears to be created by the nerve terminal itself, as crushing parts of the terminal causes a proportional loss of generator-potential amplitude. The important stimulus is a distortion of the terminal, as a general increase in pressure acting from all directions does not produce a response, while the most effective stimuli are those which cause the terminal axon to flatten in cross-section.

Figure 7.3: The Pacinian Corpuscle Generator Potential

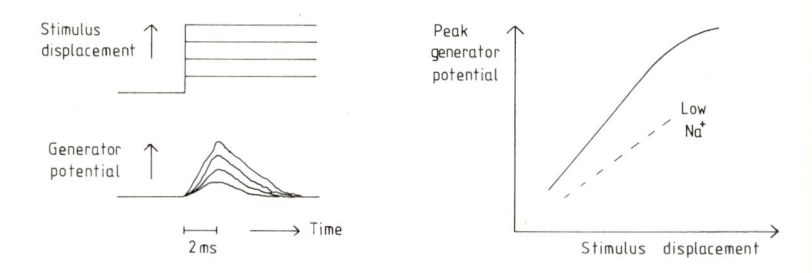

For small stimuli the amplitude of the generator potential is linearly related to the displacement of the stimulus as shown on the right of Figure 7.3, but at high displacements there is a saturation of the generator potential. Reduction of external sodium ion concentration causes a diminution in the generator potential which falls to ten per cent of its original value if all external sodium is removed. These experiments are consistent with the model illustrated in Figure 7.4, in which the generator potential results from a mechanically-induced increase in the sodium ion conductance of the nerve terminal membrane. The channels involved are distinct from those of the action potential. The saturation of the generator potential is consistent with the fact that once the membrane potential approaches V_{Na} any further opening of sodium channels will have little affect on membrane current. The postulated sequence of events causing transduction therefore involves mechanical deformation, increased membrane sodium conductance, a generator current of sodium ions into the terminal and a resultant depolarisation. Confirmation of this scheme can be obtained using the crayfish stretch receptor, from which intracellular recordings can be made. Again there is a saturable, graded depolarisation on mechanical stimulation, which is dependent on external sodium ions. Furthermore the total membrane resistance decreases during the response, and there is a reversal potential of 0 mV, indicating that the channels opened by deformation may be permeable to a number of ions and not just to sodium.

Figure 7.4: A Possible Mechanism of Mechanical Transduction Involving an Ion Channel Opened by Changes in Membrane Curvature

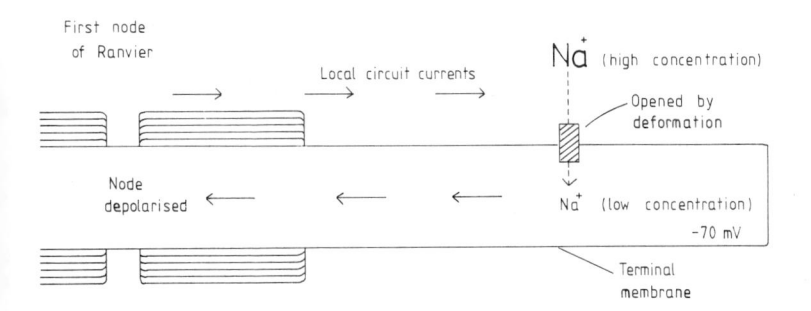

The molecular origins of the conductance changes caused by deformation of the terminal membrane are obscure. There is no known property of the lipid bilayer which could account for such large increases in conductance, and so protein channels are thought to be involved. Adrenaline has an interesting action on the pacinian corpuscle, causing an increase in the amplitude and rate of rise of the generator potential, which also suggests that transduction results from specific protein function. The generator potential is temperature-dependent; it is doubled in size by a $10°C$ rise in temperature, indicating that some chemical process takes place, possibly the movement of ions between binding sites in a protein channel. The effect of adrenaline may be of physiological importance as substances of this nature, carried by the bloodstream or locally released from nerve terminals, could be used to increase receptor sensitivity when a high level of 'attention' to stimuli is required.

7.3 The Initiation of Action Potentials

The generator potential will drive local circuit currents in the pacinian corpuscle as shown in Figure 7.4, which will tend to depolarise the first node Ranvier and initiate an action potential. It is not entirely clear whether this passive spread of current along the terminal is the whole explanation for the initiation of impulses. One problem is that most experiments employ TTX or procaine to block the action-potential mechanism, and stimuli of high amplitude are used which produce generator potentials well above the threshold for impulse production. These experiments may not be relevant to the response of the terminal to small-amplitude stimuli. It has been shown that the terminal region itself can conduct a low velocity all-or-nothing action potential, which is blocked by TTX. This phenomenon may be important in transmitting information about small-displacement stimuli from the terminal to the beginning of the myelinated section of the axon, stimuli which might otherwise not be sufficient to elicit a nerve impulse.

7.4 Adaptation in Mechanoreceptors

Figure 7.5: Mechanical Adaptation in the Pacinian Corpuscle. A — Transient Response to Intact Corpuscle. B — Sustained Response of Dissected Ending.

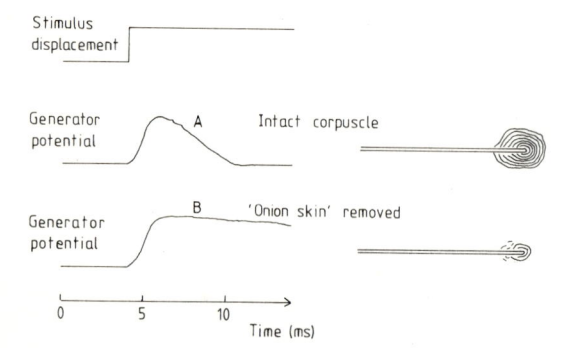

Adaptation is said to occur when a sustained stimulus to a mechanoreceptor only produces a transient nervous response. This filtering of information may occur at a mechanical level or may be due to the electrical properties of the receptor.

It was mentioned in Section 7.2 that removal of the 'onion skin' accessory structures of the pacinian corpuscle does not affect its response to a transient stimulus. This is not the case if a maintained stimulus is employed as shown in Figure 7.5. With the whole receptor only a transient generator potential is produced, while if the onion skins are removed then the generator potential is maintained for as long as the stimulus. The accessory structures produce this mechanical adaptation due to the nature of their transmission of pressure variations from the external surface of the capsule to the nerve terminal at its core. The onion-skin layers form a visco-elastic structure which transmits high frequency pressure variations, but not sustained constant displacements. The relationship between the mechanical stimulus and the generator potential of the pacinian corpuscle can be successfully explained using a mathematical model of such a visco-elastic element. When, for example, an external deformation is suddenly applied and then maintained, the core of the corpuscle undergoes only a transient deformation and then returns to its original shape. By this means only the initial application of pressure is sensed and signalled by the receptor.

Mechanical contributions to adaptation can be obviated in the pacinian corpuscle by applying direct currents to cause a sustained artificial generator potential. This does not produce an infinite train of nerve impulses of constant frequency, but elicits only a short burst of action potentials. Clearly there is also an electrical component to adaptation, and this phenomenon is known as *accommodation*. The membrane structures responsible for this phenomenon are unknown; it is not explained in any simple manner by a refractory state of the voltage-sensitive sodium ion channels of the nerve membrane.

Possessing both mechanical and electrical components, adaptation in the pacinian corpuscle is both extremely rapid and complete, so that no perturbation evokes more than one nerve impulse, and a steady-state stimulus elicits no continuous activity at all. Other receptors, notably muscle stretch receptors, show lesser degrees of adaptation to a stimulus so that their *dynamic* and *static* phases of response can be distinguished. The dynamic phase is the peak of activity due to application of the stimulus, while the static phase is the level of sustained response due to the maintained presence of the stimulus. The relative importance of the mechanical and electrical components of adaptation varies from receptor to receptor; some mammalian muscle-spindle stretch receptors show adaptation to a sustained stretch but not to a maintained depolarising current, and here mechanical effects appear to be most important.

The degree of adaptation shown by a receptor can be quantified by the ratio of its dynamic and static activities, and in mammalian muscle-spindle stretch receptors there are wide variations in this ratio. The dynamic response can be considered to be a signal of the velocity of muscle fibre shortening, while the static response is a measure of fibre length. Therefore some receptors are velocity-sensitive while others are length-sensitive and there is some evidence for an anatomical distinction between the two types. The interested reader is directed to the work of P.B.C. Mathews and P.A. Merton and colleagues on the importance of this sensory information in the control of motor function.

7.5 Summary

The mechanisms of sensory transduction in selected examples of mechanoreceptors have been described. The first stage is a mechanical transmission of pressure variations across the accessory structures of the receptor to the sensory nerve terminal. Changes in the signal take place at this stage and there may be a mechanical adaptation whereby only transient perturbations are transmitted. This filtering effect is due to the visco-elastic properties of the accessory structures. The next stage is the conversion of mechanical deformation of the nerve terminal membrane into a graded depolarisation of the terminal, called the generator potential. This is associated with a geometrically induced increase in membrane sodium ion conductance, and a resultant current of sodium ions into the terminal down their electrochemical gradient. Finally the generator potential drives local circuit current flows at the terminal, causing depolarisation at the first node of Ranvier and the initiation of nerve impulses. A form of adaptation may take place at this last stage whereby a maintained generator potential only elicits an initial burst of impulses.

The precise nature of mechanoreception varies from tissue to tissue but some points of general importance arise from these examples. Adaptation at the level of the receptor is widely seen and provides a mechanism for changes in the environment to be signalled to the CNS rather than constant influences. The phenomenon of a sodium ion-dependent nerve terminal depolarisation caused by deformation is also common, and suggests the existence of protein ion channels opened by changes in membrane geometry.

Further Reading

Catton, W.T. (1970), Mechanoreceptor function. *Physiol. Rev. 50,* 297.

Katz, B. (1950), Depolarisation of sensory terminals and the initiation of impulses in the muscle spindle. *J. Physiol. 111,* 261.

Loewenstein, W.R. (1960), Biological transducers. *Sci. Am. 203,* 98.

Loewenstein, W.R. and Mendelson M. (1965), Components of receptor adaptation in a Pacinian corpuscle. *J. Physiol. 177,* 377.

Loewenstein, W.R. and Rathkamp, R. (1958), The sites for mechano-electric conversion in a Pacinian corpuscle. *J. Gen. Physiol. 41,* 825.

Mathews, P.B.C. (1972), *Mammalian Muscle Receptors and their Central Actions.* Edward Arnold, London.

Merton, P.A. (1972), How we control the contractions of our muscles. *Sci. Am. 226,* 30.

Nakajima, S. and Odenera, K. (1969), Membrane potentials of the stretch receptor neurones of crayfish with particular reference to mechanisms of sensory adaptation. *J. Physiol. 200,* 161.

Part Four

INFORMATION INTO ACTION

8 SKELETAL MUSCLE ACTIVATION

8.1 Introduction

Voluntary movement is accomplished by cylindrical contractile cells known as skeletal or striated muscle fibres, which are controlled by motor nerves originating in the central nervous system. As shown in Figure 8.1 the muscle cell receives a synaptic input at its endplate region, capable of evoking a contraction in the whole fibre. Each motor nerve may branch extensively thus synapsing onto a number of muscle cells and activating them all synchronously so they function as a *motor unit*. It was described in Chapter 5 how the nerve-muscle synapse at the endplate allows a nerve impulse to cause a local depolarisation of the muscle cell membrane. The object of this chapter is to explain how this local excitation is spread rapidly to all parts of the fibre, and then coupled to the contractile machinery to cause muscle shortening. It will be seen that this process of excitation-contraction coupling involves a specialised network of excitable membranes within the muscle cell itself.

Figure 8.1: The Skeletal Muscle Cell. Arrows represent the spread of excitation across the cylindrical surface. Spread must also occur into the cell interior.

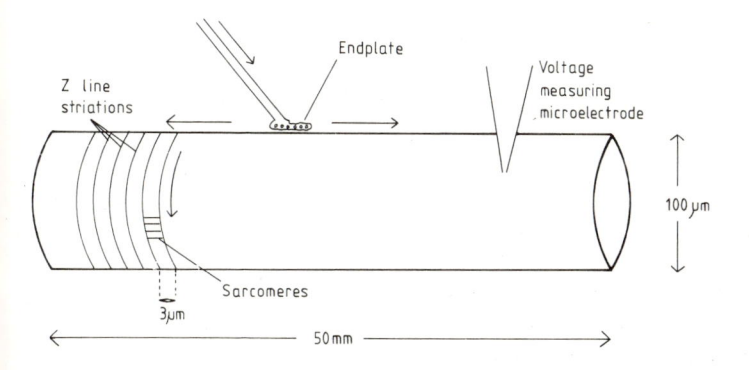

In a muscle fibre the functional unit of contraction is the *sarcomere,* which is about 3 μm in length, and each sarcomere must be individually activated. Accordingly there are two dimensions in which the spread of muscle excitation must take place: longitudinally to the ends of the fibre and radially into the core of the cylinder from the surface regions. The muscle can only exert force efficiently if all of its parts are activated within a few milliseconds of each other. The longitudinal spread is achieved by a nerve-like action potential travelling along the surface membrane of the fibre, while the radial spread is due to the transverse tubules (invaginations of the surface membrane) which pass across the whole diameter of the cell. The electrical activity in these membrane structures causes a release of calcium ions from local intracellular stores, which activates the contractile proteins of the muscle. These proteins generate tension by a sliding filament mechanism (which will not be discussed here) in which contraction is controlled by the local concentration of free calcium ions. For a treatment of the sliding filament theory the review by H.E. Huxley (1969) is recommended.

The response of a skeletal muscle fibre to a nerve impulse is to give a single twitch of contraction which is always of considerably longer duration than the muscle action potential. A train of impulses can elicit a summation of twitches and a maintained tension in the fibre. Different skeletal muscles can be classified as fast or slow according to the duration of their single twitch but the basic mechanisms of response are essentially similar, and the data used in the following discussion have been obtained mainly from fast muscle fibres in the frog.

8.2 The Structure of a Skeletal Muscle Cell

The structure of skeletal muscle has been investigated with light microscopy, electron microscopy and X-ray diffraction techniques. The muscle fibre is a cylindrical multinucleate cell containing an orientated filamentous ultrastructure due to the contractile proteins actin and myosin. Dark lines across the cylinder known as Z lines mark the ends of each sarcomere and form the striated appearance which gives these fibres their descriptive name. Activation of the sliding filaments causes them to pull the Z lines closer together and shorten the whole cell. The structure is schematically drawn in Figure 8.2, where the internal membrane system is also represented. The transverse tubules (TT) are cylindrical invaginations of the surface membrane which pass right across the fibre and contain extracellular fluid. In the frog these tubules occur at the level of each Z line and so there is one at the end of each sarcomere. The sarcoplasmic reticulum (SR) is an internal membrane structure which encloses its own space within the muscle. It comprises terminal cisternae which lie close to the transverse tubules and are connected to a longitudinal reticulum near the centre of the sarcomere. There is a region of close apposition of transverse tubular and sarcoplasmic reticular membranes which is held to be of great importance in the functional coupling of the two systems. The membranes lie about 12 nm apart and there are a series of electron dense 'feet' projecting from the sarcoplasmic reticulum, at intervals of about 30 nm, which approach to within 5 nm of the transverse tubular membrane, and which are joined to it by an amorphous material of unknown composition.

Figure 8.2: The Ultrastructure of a Skeletal Muscle Cell. Inward spread of activation occurs down the transverse tubules.

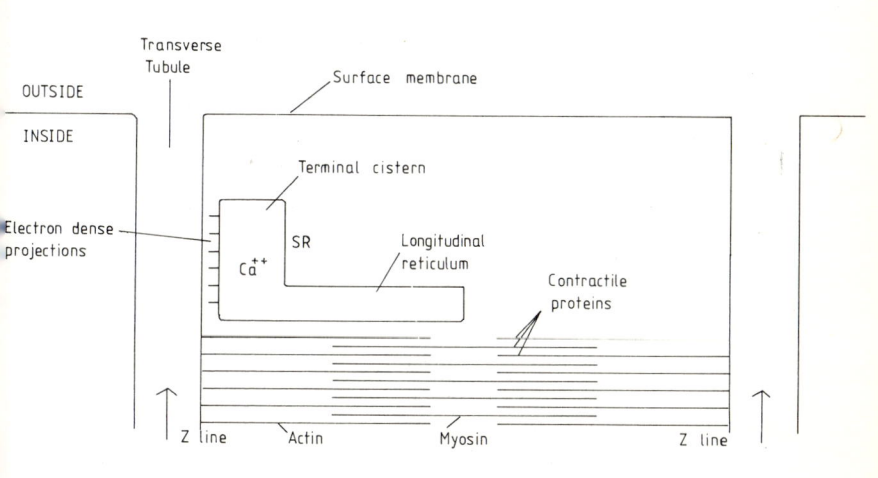

Continuity between the transverse tubular and surface membranes has been established by showing that the electron-dense protein *ferritin* with a diameter of 11 nm will diffuse from the external medium into the transverse tubules, but not into other internal spaces. Furthermore the electrical capacity of the muscle cell membrane is found to be between 6 and 8 μF per square centimetre of geometric surface, compared to a value of 1 μF per square centimetre in the squid axon. This difference

102 Skeletal Muscle Activation

can be largely explained by taking into account the contribution made by the transverse tubules connected to the surface membrane. Treatment of a muscle fibre with glycerol severs the connection between the surface and transverse tubular membranes and reduces the electrical capacity as expected. The sarcoplasmic reticulum is the site of the intracellular store of calcium ions which are used to initiate muscle contraction. Some evidence for this can be obtained using the technique of autoradiography in which the muscle fibre is loaded with radioactive $^{45}Ca^{++}$, and after fixation will produce an image of its own calcium ion distribution on a photographic plate. The calcium is found to be concentrated in the terminal cisternae of the sarcoplasmic reticulum.

8.3 The Muscle Action Potential and Transverse Tubular Spread

Figure 8.3: A — The Skeletal Muscle Action Potential. B — The Muscle Tension Produced by Artificial Membrane Depolarisation.

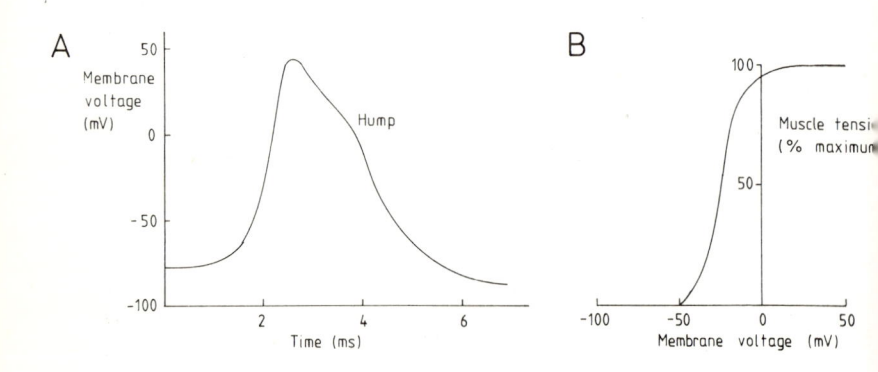

Sodium and potassium ions are distributed across the muscle membrane in the usual manner for an excitable cell, with sodium at a high concentration outside and potassium at high levels inside. The chloride ion is present at higher concentrations outside than inside and the resting potential of the muscle is −85 mV which is close to the Nernst potentials both for potassium and chloride. As expected, the resting membrane shows higher conductances to chloride and potassium than to sodium ions. On depolarisation the muscle membrane exhibits a propagated action potential, which can be monitored using intracellular electrodes and is illustrated in Figure 8.3A. It is similar to the nerve impulse in being a transient reversal of membrane potential to positive values, and it utilises the same voltage-sensitive sodium and potassium channels as are found in the nerve membrane. The muscle impulse has a characteristic shape with a hump on the repolarising phase which is due to current flow into the transverse tubules. It is conducted along the fibre with a velocity of about four metres per second. Tetrodotoxin (TTX) abolishes the impulse and thereby inhibits the longitudinal spread of muscle activation, but a fibre treated with TTX will still contract locally if depolarised by an electrode. Therefore the depolarisation of membrane voltage associated with passage of the action potential is held to be the signal for contraction. Further evidence for this view is that rapidly increasing the extracellular potassium ion concentration, to depolarise the membrane, also causes a contractile response.

The relationship between surface membrane depolarisation and contraction is represented in Figure 8.3B, which shows that tension increases rapidly as membrane potential changes from -50 to -10 mV and then reaches a maximum value. If depolarisation is artificially applied and maintained (using external potassium ions) then the tension is not constant but shows inactivation over a period of about one second.

In normal conditions the filaments of the whole cross-section of the muscle cylinder contract almost simultaneously despite the fact that the depolarising impulse travels down the surface membrane. This rapid inward radial spread of information is achieved by the transverse tubular system. Huxley and Taylor (1958) used local extracellular stimulating currents on frog muscle and found that inward spread of a small stimulus would only occur if it was applied opposite a Z line and the associated transverse tubule. In lizard muscle the Z lines and transverse tubules are not coincident, but again the transverse tubule is the most sensitive site for radial spread. In all skeletal muscles de-tubulation with glycerol blocks the inward spread of small stimuli. The depolarisation of the surface membrane is conducted along the transverse tubule by a form of active propagation akin to an action potential turned inside-out. The inward spread of contraction can be watched microscopically and is found to be TTX sensitive, sodium dependent and increased in rate by a factor of 2.3 on raising the temperature by 10^{0}C; at 20^{0}C the rate of radial conduction is seven centimetres per second. In the presence of TTX increasing levels of depolarisation at the surface membrane will always elicit contraction at the fibre surface before contraction of the axial filaments, but in the absence of TTX an axial contraction can be seen with stimuli which produce no surface response. This indicates an action potential-like spread of depolarisation along the transverse tubule rather than a passive flow of current. Adrian and Peachey (1973) demonstrated that the complex form of the muscle action potential could be well represented using a mathematical model in which the transverse tubular membrane contains voltage-sensitive sodium and potassium channels at one-twentieth of their density in the surface membrane.

The propagation of an 'inside-out' action potential down the transverse tubules involves a flux of sodium ions into the cytoplasm from the tubular lumen and a flux of potassium ions in the reverse direction. This loss of potassium into the tubule is minimised by two factors: first the high chloride ion conductance of the surface membrane which allows chloride influx to dissipate depolarisation, and secondly the resting potassium permeability channel shows 'inward rectification' by which it allows potassium ions to enter the cell more easily than they may leave. If there is a defect in these ionic conductance systems of the muscle membrane then a significant accumulation of potassium in the transverse tubules may occur with clinical consequences as, for example, in the disease of *myotonia* discussed in Chapter 12.

8.4 Sarcoplasmic Recticulum and Calcium Ion Release

Having spread the signal for muscle activation to each sarcomere by depolarisation of its local transverse tubular membrane, the next step in excitation-contraction coupling is the transmission of information from the transverse tubule to the contractile proteins actin and myosin. There is strong evidence that the final signal to these proteins is the release of calcium ions from internal stores in the sarcoplasmic reticulum. Direct injection of calcium ions through a micropipette electrode will evoke a local contraction in the absence of electrical activity. The dye *aequorin* fluoresces in the presence of calcium ions and can be used to confirm that calcium ions are released in physiological contraction as shown in Figure 8.4. After injection of aequorin into the cell a fluorescent response is seen to follow the electrical response and precede the contraction. The resting intracellular concentration of free calcium ions in aqueous solution can be estimated, using a calcium ion binding agent (EGTA) and the aequorin response, to be about $0.1 \mu M$. During normal muscle activation this concentration rises to about 10 μM and then rapidly returns to the resting value. Using EGTA to buffer the calcium ions it is found that a calcium concentration of about 0.5 μM is the threshold for a contractile response. These estimates are consistent with the known calcium ion binding properties of the protein troponin which regulates the force-generating activity of the actin and myosin filaments.

Figure 8.4: The Aequorin Response as Evidence that a Rise in Internal Calcium Ion Concentration is a Step in Excitation-contraction Coupling

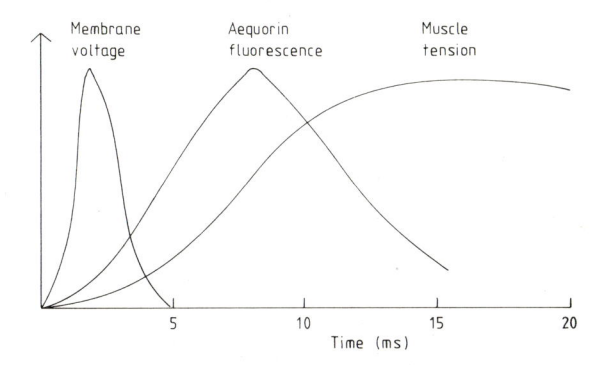

The next question which arises concerns the origin of the calcium ions mobilised on excitation. Extracellular calcium ions are unimportant for contraction as normal, stimulated shortening occurs even after EGTA has been used to reduce external calcium ion concentration to below 0.01 μM. The total fibre content of calcium is large and equivalent to about 1.0 mM if the ions were uniformly distributed. Autoradiographic studies indicate that these ions are concentrated in the sarcoplasmic reticulum (SR) which represents approximately ten per cent of cell volume, and so must contain calcium at a total level of 10 mM. This idea of a role for the SR in calcium storage is supported by experiments on isolated SR vesicles. The vesicle membrane possesses a protein pump which uses ATP as an energy source to transport calcium ions into the SR. This pump has an affinity for calcium ions high enough to explain the resting cytoplasmic concentration of 0.1 μM, and it can create a transmembrane calcium ion concentration gradient of three thousand-fold. The pump is therefore capable of creating a free calcium ion concentration in the SR of 300 μM. In addition the SR contains large amounts of a calcium-binding protein (CBP or calsequestrin) which allows the total store of calcium in the SR to approach the estimated concentration of 10 mM, as over 95 per cent of calcium in the SR is protein-bound. This distribution of calcium ions in a muscle cell is illustrated in Figure 8.5. The calcium-binding protein (CBP) and the calcium pump function together to produce an extremely low cytoplasmic calcium ion concentration with large stores in the SR.

Figure 8.5: The Distribution of Calcium is a Muscle Cell Between SR and Sarcoplasm

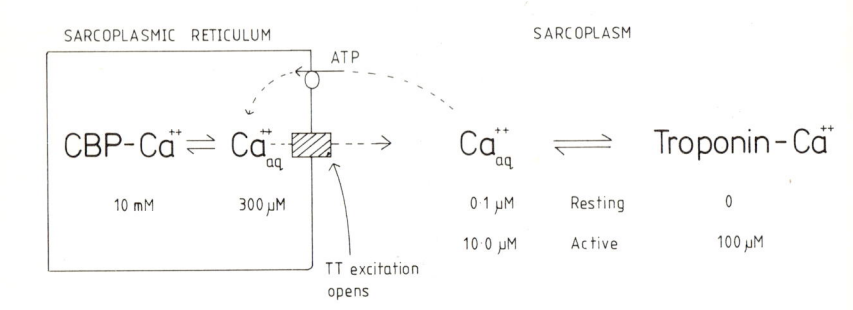

SARCOPLASMIC RETICULUM		SARCOPLASM	
$CBP\text{-}Ca^{++} \rightleftharpoons Ca^{++}_{aq}$	$\xrightarrow{ATP}$	$Ca^{++}_{aq} \rightleftharpoons$	$Troponin\text{-}Ca^{++}$
10 mM	300 μM	0.1 μM Resting	0
		10.0 μM Active	100 μM

TT excitation opens

On excitation there is a large release of calcium from the SR into the cytoplasm, and these ions almost all bind to troponin molecules and activate contraction. Only a small percentage of released calcium ions actually remain in the cytoplasm, and increase the concentration there from 0.1 to 10 μM. The membrane of the SR appears to respond to transient depolarisation of the nearby transverse tubular membrane with a transient increase in its calcium ion conductance. The precise form of this response is difficult to study as calcium efflux is driven by an electrochemical gradient, not simply a chemical gradient, and the voltage across the SR membrane is unknown. The rise in cytoplasmic calcium ion concentration is transient because release is not maintained, and because ion pumping back into the SR is rapid. There is autoradiographic evidence that calcium release takes place from the terminal cisternae of the SR close to the transverse tubules, while re-uptake occurs into the longitudinal reticulum. This indicates a cycling of calcium during muscle activity, including transport within the sarcoplasmic reticulum itself.

8.5 Molecular Mechanisms of Coupling between TT and SR

The mechanism of information transfer between TT and SR membranes is a problem which has stimulated considerable recent discussion and research. Attention has naturally focused on the regions of close apposition of the two membrane systems where the characteristic dense projections are seen in electron micrographs, although no precise function has yet been ascribed to these structures. Three possible mechanisms of functional coupling may be distinguished as the chemical, electrical and mechanical hypotheses.

The *chemical hypothesis* is that a release of calcium ions from the transverse tubule, into the narrow gap between the membranes induces calcium ion release from the SR in a self-regenerative manner. Here, internal release is dependent on a small influx of external calcium. The evidence for this model is that the sarcoplasmic reticulum can be made to release calcium by the direct application of calcium to its membrane. The physiological relevance of this calcium-induced calcium release is questionable as concentrations as high as 0.1 mM are needed, far greater than the normal cytoplasmic calcium ion concentration, even during conctraction. Further problems are that this form of release is only seen if the sarcoplasmic reticulum is pre-loaded with calcium, and that procaine blocks this release but not normally-stimulated release. Finally this hypothesis would predict that removing extracellular calcium ions should affect the contractile response, which is not the case.

The *electrical hypothesis* is that transverse tubular depolarisation causes depolarisation of the SR membrane, which as a result undergoes an increase in calcium ion permeability. There is indirect evidence for an electrical coupling between transverse tubule and SR. Sucrose applied externally to a muscle fibre causes swelling of the SR, as if the lumen of the SR was continuous with the extracellular space. Certain fluorescent dyes respond to muscle excitation in a manner suggesting that a large change in the membrane potential of the SR takes place. Furthermore direct depolarisation of the SR will evoke calcium release. However it is not possible to quantify these voltage changes across the SR membrane, and the electrical effects during muscle excitation could be the result of calcium release rather than the cause of it. Also voltage-clamp studies on muscle have not detected a component of current flow ascribable to the SR, and this result argues against an electrical coupling.

The *mechanical hypothesis* of transverse tubule-SR coupling is the most recent. It suggests that molecular movements of charge within the transverse tubular membrane, caused by voltage changes, are mechanically coupled to a gate on a calcium ion channel of the SR membrane 12 nm away. The model is illustrated in Figure 8.6. When the transverse tubular membrane depolarises a set of positive charges moves across it towards the lumen of the TT. These charges are attached to a structure joining the two membranes, which at rest is blocking a calcium channel in the SR, and which moves with the charges to open the channel. By this mechanism, depolarisation of the transverse tubular membrane is coupled to the calcium ion conductance of the SR. The major evidence for this model is the detection within the muscle membrane system of charge movements, which show the correct dependence on membrane voltage to explain the voltage-tension relationship of Figure 8.3B. These charge movements were first detected by Schneider and Chandler (1973) and are inhibited by glycerol detubulation of the fibre. If depolarisation is maintained the charge movement is inactivated on a time-scale consistent with the inactivation of contraction. This could be due to a cage of negative charges within the transverse tubular membrane which moves slowly to pinion the positive charges back in the 'channel closed' position (Figure 8.6C). After inactivation when the normal resting potential is restored there is a repriming of charge movement which shows the same time-course as the repriming of the contractile response.

It is tempting to associate these transverse tubular charge movements with the electron-dense feet projecting between the transverse tubules and the SR. There is agreement between the number of sites of charge movement and the surface density of these feet, which is estimated to be 500 per square micron. If these structures are the sites of calcium ion efflux then each 'foot' must conduct about 300 ions in three milliseconds. This is a flux rate of 100,000 ions per second which is easily achievable by the pore type of ion channel found in other excitable membranes. However any conclusions about the structure of an SR calcium ion channel would be premature. Indeed such a channel may prove very difficult to investigate because the correct alignment of TT and SR could be essential for its function and so extracted vesicles of SR would not be experimentally useful.

In conclusion the precise mechanism of TT-SR coupling remains obscure, although there are good reasons to believe that voltage-dependent charge movements within the transverse tubular membrane are important in the activation of contraction.

Figure 8.6: A Mechanical Hypothesis of Coupling Between TT and SR. A — Resting state with SR channel blocked. B — TT depolarisation moves charges and opens channel. C — Cage of negative charges inactivates channel.

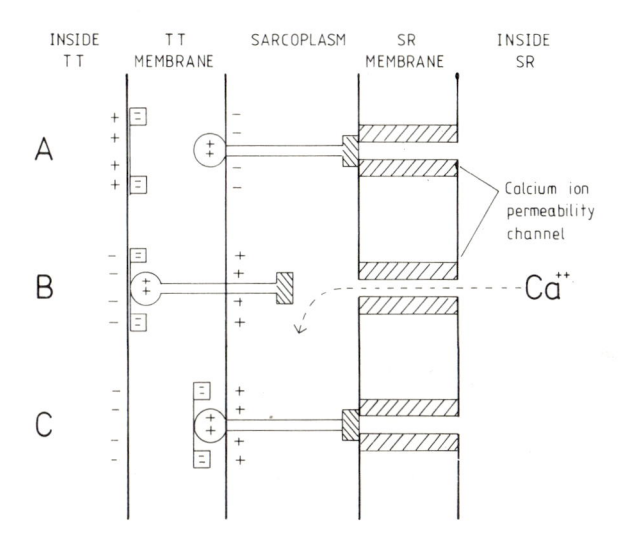

8.6 Summary

In this chapter the process of excitation-contraction coupling in skeletal muscle has been described. This is the mechanism by which a local depolarisation of the endplate region of a skeletal muscle cell causes contraction in all regions of the cell, as illustrated in Figure 8.7. It begins with the initiation of a muscle action potential at the endplate due to depolarisation. This action potential is propagated along the surface membrane to the ends of the cylindrical muscle cell, and the depolarisation associated with this impulse is the signal for contraction. An inward spread of depolarisation occurs by active spread into the transverse tubular system, so that a voltage change occurs across transverse tubular membranes deep in the cell. Each small sarcomere of contractile filaments is excited by the depolarisation of its local transverse tubular membrane. This TT depolarisation produces an increase in the calcium permeability of the nearby sarcoplasmic reticulum and a release of calcium ions from stores in the SR. The contractile machinery includes a regulatory protein component called troponin which

110 Skeletal Muscle Activation

binds these released calcium ions, and by doing so induces conformational changes in the filaments and initiates contraction. The transient depolarisation associated with a single muscle action potential produces only a transient release of calcium ions which are rapidly pumped back into the sarcoplasmic reticulum, so that the contractile response is a short-lived twitch. The coupling between the membranes of the transverse tubule and sarcoplasmic reticulum is not well understood but may involve a movement of charges within the transverse tubular membrane itself.

Figure 8.7: A Summary of Excitation —contraction Coupling in Skeletal Muscle

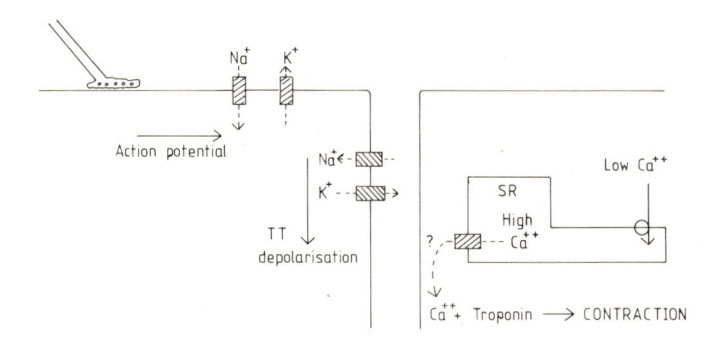

In the next two chapters cardiac muscle and smooth muscle will be described. These muscles differ functionally from skeletal muscle and also exhibit characteristic differences in the manner and form of their activation. The following features of skeletal muscle are of interest for comparative purposes and show how its excitable characteristics are related to its function.

(1) In the absence of nerve stimulation a skeletal muscle cell produces no tension. It only contracts when evoked by activity in its motor nerve. There is no electrical activity at rest.
(2) The twitch response of a skeletal muscle to a single impulse is of long duration compared to the time-course of the muscle action potential. Therefore on repetitive stimulation, the muscle cell will exert a maintained tension.
(3) There is an extensive transverse tubular system for rapid

activation of the whole volume of the skeletal muscle cell.
(4) Calcium ions for initiation of contraction are released from internal stores close to their site of action on the contractile proteins. Entry of calcium from the external medium is not important.
(5) There is no direct electrical coupling between cells.

Further Reading

Adrian, R.H. and Peachey, L.D. (1973), Reconstruction of the action potential of frog sartorius muscle. *J. Physiol. 235*, 103.

Chandler, W.K, Rakowski, R.F. and Schneider, M.F. (1976), Effects of glycerol treatment and maintained depolarisation on charge movements in skeletal muscles. *J. Physiol. 254*, 285.

Constantin, L.L. (1970), The role of sodium current in the radial spread of contraction in frog muscle fibres. *J. Gen. Physiol. 55*, 703.

— (1975), Contraction activation in skeletal muscle. *Prog. Biophys. Mol. Biol. 29*, 197.

Ebashi, S. (1976), Excitation-contraction coupling. *Ann. Rev. Physiol. 38*, 293.

Endo, M. (1977), Calcium release from the sarcoplasmic reticulum. *Physiol. Rev. 57*, 71.

Fuchs, F. (1974), Striated muscle. *Ann. Rev. Physiol. 36*, 461.

Gonzalez-Serratos, H. (1971), Inward spread of activation in vertebrate muscle fibres. *J. Physiol. 212*, 777.

Huxley, A.F. (1971), The activation of striated muscle and its mechanical response. *Proc. Roy. Soc. London B. 178*, 1.

Huxley, A.F. and Taylor, R.E. (1958), Local activation of striated muscle fibres. *J. Physiol. 144*, 426.

Huxley, H.E. (1969), The mechanism of muscle contraction. *Science 164*, 1356.

Rudel, R. and Taylor, S.R. (1973), Aequorin luminescence during contraction of amphibian skeletal muscle. *J. Physiol. 233*, 5P.

Schneider, M.F. and Chandler, W.K. (1973), Voltage dependent charge movement in skeletal muscle: a possible step in excitation-contraction coupling. *Nature 242*, 244.

9 CARDIAC MUSCLE

9.1 The Heart as a Pump

The heart is a muscular organ which pumps blood around two circulations. The left side of the heart receives oxygenated blood from the lungs, and pumps it into the arteries of the systemic circulation to supply the tissues of the body. The right side of the heart receives deoxygenated venous blood from the tissues, and pumps it through the lungs in the pulmonary circulation. Each of these two pumps consists of two chambers, an atrium and a ventricle. The thick walls of these chambers are formed by cardiac muscle. Rhythmic contraction of this muscle provides the force needed to perform the work of pumping.

Cardiac muscle has a number of special properties which are adapted for efficient pumping. Many of these are based on the ionic current flows across the cardiac muscle cell membrane. One important property is that cardiac muscle exhibits automatic rhythmic electrical activity, and will continue to fire without any external stimulus. The rate of this activity varies from one region of the heart to another, and the fastest rhythm occurs in the area of the right atrium known as the sino-atrial (SA) node. Electrical activity spreads from cell to cell across the cardiac muscle mass, and through specialised conducting tissue, so the SA node is able to drive the whole heart at its own rate. It is known as the *pacemaker* region and it sets the heart rate in man about 70 beats per minute. In practice, therefore, the potential autorhythmicity of each myocardial cell is not utilised. The excitation of all cardiac muscle cells, except those of the SA node, is stimulated by current flow from neighbouring cells.

The blood flowing into the heart passes first into the atria. Atrial contraction is weak and serves mainly to complete the filling of the ventricles. Ventricular contraction is much more powerful and expels the blood from the heart. Atrial contraction immediately precedes ventricular contraction so that ventricular filling is completed at the correct moment. This sequence is achieved by the specialised conducting system within the heart which spreads the excitation initiated by the SA node. At rest in man the volume of blood expelled per beat is about 70 ml. Therefore the resting cardiac output is about five litres per minute.

Ion-flows across the cardiac muscle cell membrane are responsible for the generation and spread of electrical activity, as well as for the coupling of electrical and mechanical excitation. Heart rate and force of contraction are controlled by nerves which synapse on the cardiac muscle and influence these ionic currents. The cardiac ionic currents can be detected as changes in the electrical potential between electrodes on the skin, and recorded as the electrocardiogram (ECG). The ECG is an important diagnostic tool in clinical medicine as heart disease is often accompanied by characteristic changes in the cardiac currents. The purpose of this chapter is to describe the excitable properties of cardiac cells and relate them to cardiac muscle function.

9.2 The Structure and Conducting System of the Heart

Cardiac muscle cells are complex in structure. Each cell branches and forms junctions with neighbouring cells to produce a network of interconnected fibres as shown in Figure 9.1A. The intercalated discs at the cell junctions are functionally significant, as a part of the disc is a low resistance pathway allowing direct electrical coupling between the cells. By these pathways electrical excitation of one cardiac cell will spread to its neighbours and eventually to the whole heart. The low-resistance pathways allow ions of considerable size to pass: the dye procion yellow (molecular wt 700) will diffuse from one cell to another. Within the heart there are groups of cells which provide specialised pathways for the spread of excitation, as illustrated in Figure 9.1B. These pathways are known as the conducting system of the heart. The SA node pacemaker region with the highest intrinsic rate of firing initiates excitation which spreads out across the right and left atria. Spread to the ventricles does not immediately occur, as the atrio-ventricular (AV) septum is a barrier to current flow. The ventricles are excited by a specific path which begins at the atrio-ventricular (AV) node, continues down the bundle of His, and then branches to all parts of the ventricles in the Purkinje fibres. The times for conduction from the SA node to each area of the heart (in milliseconds) are shown in Figure 9.1B. A delay of about 100 ms occurs at the AV node, but after this, spread down the bundle of His and Purkinje fibres is rapid. The delay ensures that the atria have time to pump blood into the ventricles before ventricular contraction, so that the cardiac output of blood at each beat is maximal. Contraction is almost synchronous in different regions of the ventricles as a result of fast

Purkinje fibre conduction, and this also contributes to the efficiency of the pump.

Figure 9.1: A — Cardiac Muscle Cells. B — The Conducting System of the Heart. The time taken (in milliseconds) for excitation to travel from the SA node is indicated by the figures.

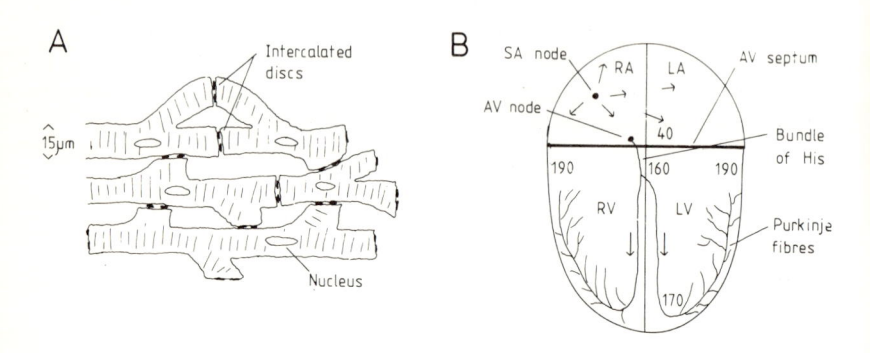

After one wave of cardiac contraction excited by the SA node there is a period of quiescence, during which refilling of the atria and ventricles with blood takes place. Then the SA node fires again and the cycle is repeated. The node therefore controls the cardiac contraction as long as it is the fastest pacemaker, and as long as the conducting system is intact. In certain pathological states these criteria no longer apply and the heart beat shows abnormal rate and rhythm with a loss of pumping capacity. In the normal cycle the period of excitation and contraction is known as *systole* while the period of relaxation is known as *diastole*.

Cardiac muscle cells from different parts of the heart differ in their electrical and mechanical properties. For example Purkinje fibres show fast electrical conduction but are not contractile. Atrial cells conduct electrical activity more slowly and are contractile. All cardiac cells are capable of automatic rhythmic electrical activity, but the rate of such activity varies from area to area. The SA node shows the fastest rhythm, while the AV node, Purkinje fibres and ventricular muscle cells fire progressively more slowly when isolated from the pacemaker under experimental conditions. The internal cellular structure of cardiac muscle shows considerable similarities to skeletal muscle. There are protein filaments of actin and myosin, the contractile machinery, with a striated appearance. An internal membrane system of transverse tubules and sarcoplasmic reticulum exists, but is not as extensive as in skeletal muscle. The sarcoplasmic reticulum, as in skeletal muscle, is a store of calcium ions.

9.3 The Cardiac Action Potential

Figure 9.2: Cardiac Action Potentials

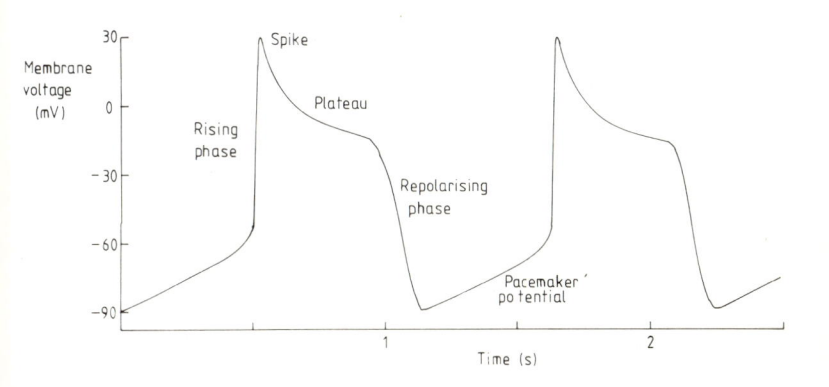

The intrinsic electrical activity of cardiac muscle can be recorded using an intracellular electrode. Figure 9.2 is a diagram of cardiac action potentials measured in a sheep Purkinje fibre. These action potentials show characteristic features which distinguish them from the nerve action potential of Chapter 4. There is still a rapid rising phase which overshoots zero to take membrane voltage to positive values. After a slight decline from this initial spike, there is a long plateau region before a slow repolarisation back to negative potentials. There is then a gradual rise in potential which leads to the initiation of another spike. This gradual depolarisation between -90 and -60 mV is the *pacemaker potential* which ensures that there is a spontaneous train of action potentials. As a result of the plateau the action potential is very long and lasts about 400 milliseconds. The contraction of cardiac muscle (systole) occurs during the action potential while diastole occurs during the pacemaker potential. Each contractile response lasts only as long as the action potential. Because the action potential is long-lasting, so also is the electrical refractory period, and no new action potential can occur before completion of the cardiac contraction. Heart muscle will consequently never maintain a contraction and produce a sustained tension, but can only contract rhythmically, and this is clearly consistent with its function as a pump.

The precise form of the cardiac action potential varies from region to region of the heart. The pacemaker potential at the SA node rises from −60 to −30 mV, so the membrane potential is less negative at this stage than in the Purkinje fibre. Associated with this is a more rapid rate of rise of the pacemaker potential, leading to a shorter time period between action potentials. The SA node action potential also has a shorter plateau. These two effects explain the higher intrinsic firing rate of the SA node region. In normal conditions the SA node pacemaker drives the rest of the heart. Thus other cardiac muscle cells are induced to fire not by their intrinsic pacemaker activity, but by current flows from neighbouring cells. Only in pathological conditions, for example if the AV node is blocked, do the pacemaker activities of other regions of the heart come into play. In AV block the Purkinje fibres might pace the ventricles at about 40 beats per minute.

Table 9.1: Typical Ionic Concentrations in Cardiac Muscle

Ion	Intracellular concentration (mM)	Extracellular concentration (mM)	Nernst membrane potential (mV)
Sodium	30	140	+41
Potassium	140	4	−94
Calcium	0.0001	2	+133

Table 9.1 shows the approximate concentrations of sodium, potassium and calcium ions inside and outside cardiac cells. The Nernst equilibrium potentials for each of these ions, calculated by Equation 3.5, are also given. At all times during the cardiac action potential the electrochemical driving force on sodium and calcium ions is inward while that for potassium is outward. The action potential results from currents across the cardiac muscle cell membrane carried by these ions. These currents also cause the spread of electrical activity within the heart and the initiation of the mechanical contractile response.

9.4 Ionic Currents in Cardiac Muscle

The investigation of ionic currents in cardiac muscle presents greater difficulties than in the nerve axon. Among the reasons for this are the complex geometry of the cardiac cell, the electrical coupling to neighbouring cells, and the small extracellular spaces which present a high resistance to current flow. The technique of voltage clamping is feasible in cardiac muscle, but its results are far less precise than in the nerve. Nevertheless the method is fundamental to our present understanding of cardiac ion flows and of the conductance changes of cardiac cell membranes during excitation.

The fast, rising phase of the cardiac action potential is due to sodium ion entry. The spike is lost if sodium is removed from the external solution or if tetrodotoxin (TTX) is applied to the cell. The cell membrane undergoes a voltage-dependent increase in sodium ion conductance which is self-generating, and appears to be due to gated sodium channels similar to those found in nerve membranes. As in nerve cells the sodium current inactivates at positive membrane voltages so that the initial phase of the action potential is a transient spike. After inactivation the sodium channel is refractory and it 'reprimes' at negative potentials to a viable state. In nerve, repriming is rapid and completed in a few milliseconds. A significant difference in cardiac muscle is that repriming of the sodium channel is very slow and may take as long as 500 milliseconds. This is important in ensuring that the cardiac action potential has a long refractory period; no new spike is possible until the sodium channels are functioning again. Large currents flow during the fast sodium spike and they are instrumental in the rapid excitation of neighbouring cells. The most rapidly-conducting regions of the heart have the largest TTX-sensitive current flows, while in other areas this current is small. For example the Purkinje fibres have a significant spike and a conduction velocity of about three metres per second, while the delaying cells of the AV node, which show little rapid inward current, have conduction velocities around 0.1 metres per second.

The plateau region of the cardiac action potential represents a low-conductance state of the membrane, during which a small inward current of calcium ions is balanced by a small outward current of potassium ions. The inward calcium current can be investigated using the voltage-clamp, after inhibiting the fast sodium current with TTX. Calcium influx is slowly activated by depolarisation and slowly inactivated, in contrast to the rapid response of the sodium current. That the current is carried by calcium ions is demonstrated by removing external calcium to inhibit it, and by finding a reversal potential near V_{Ca}. There appears to be a voltage-sensitive calcium ion channel in the cardiac cell membrane. It is blocked by manganese ions and by the drugs verapamil and D600, but it will pass strontium and barium ions instead of calcium. The flux of calcium into the cardiac muscle cells during excitation is important in the initiation of contraction. Calcium current inactivation may be one of the factors determining the time of repolarisation and hence the length of the plateau and action potential.

The major changes in membrane permeability during the repolarisation of the action potential and during the pacemaker potential are due to alterations in membrane potassium ion conductance. As the Nernst potential for potassium is -94 mV the net passive flux of this ion is always outwards. At the end of the plateau potassium conductance increases and the outward current causes repolarisation. During the pacemaker potential there is a small influx of sodium ions and efflux of potassium. The potassium conductance falls during this period, and the decreasing potassium current allows membrane voltage to drift upwards away from V_K toward the threshold for the next action potential. The magnitude and rate of change of the potassium current determine the rate of rise of the pacemaker potential. These voltage and time-dependent changes in potassium conductance are due to at least two different potassium channels. Neither of these is like the potassium channels of the nerve membrane. Analysis of these channels is complicated by the fact that different areas of the heart show different electrical properties during the pacemaker potential. For a fuller discussion of these potassium channels the reader is referred to the book by Noble (1979) and the review by Vassalle (1979).

9.5 The Initiation of Cardiac Contraction

The slow inward calcium ion current during the plateau region of the cardiac action potential is an essential step in the activation of contraction. Unlike skeletal muscle, removal of external calcium ions inhibits cardiac contraction. The magnitude of inward calcium current is a function of membrane voltage, as shown in Figure 9.3. This is of the same form as the relationship between membrane voltage and muscle tension. Addition of manganese ions or verapamil to inhibit calcium currents also reduces the tension, while adrenaline increases both the calcium current and the tension. Confirming the importance of calcium entry Allen and Blinks (1978) have detected an increase in internal free calcium ion concentration preceding contraction in amphibian cardiac cells, using the aequorin response.

Figure 9.3: Calcium Current and Cardiac Muscle Tension as a Function of Membrane Voltage. Adrenaline increases both.

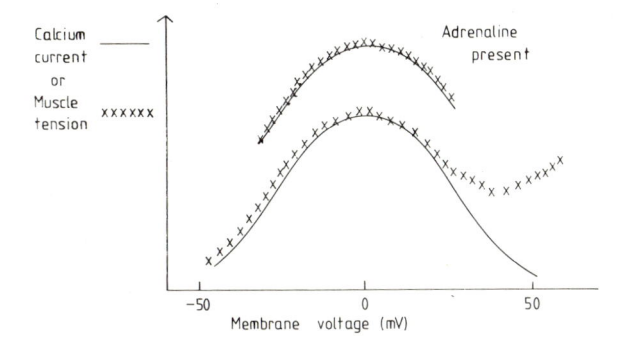

The contractile proteins of cardiac muscle are activated by calcium ions in the same manner as in skeletal muscle. The magnitude of calcium ion influx, however, is not sufficient to explain fully the contractile response. For example at positive membrane potentials the calcium ion current falls as shown in Figure 9.3 but the tension is maintained. There is also a 'staircase effect' whereby the contractile response to each of a train of action potentials becomes larger and larger, although the calcium ion current during each of them is the same. Other mechanisms for increasing free internal calcium ion concentration must exist. One source of calcium is the sarcoplasmic reticulum (SR) of cardiac muscle, which may release calcium when the transverse tubules are depolarised, as in skeletal muscle. The SR also shows calcium release stimulated by calcium itself. This calcium-induced calcium release would enable the influx of external calcium to trigger further release from internal stores. Another possibility is that influx of calcium across the cell membrane may occur by more than one route. An exchange protein exists in the cardiac membrane which will pass two sodium ions one way, if it can simultaneously transport one calcium ion in the opposite direction. This may be responsible for calcium entry without producing detectable electrical effects. These possible mechanisms of internal calcium ion concentration changes on cardiac excitation are summarised in Figure 9.4. After contraction, the relaxation of cardiac muscle is accomplished by binding of calcium to the SR and mitochondria. In the long term there is an energy-dependent extrusion of calcium ions from the cardiac cell, to maintain the electrochemical gradient.

Figure 9.4: Possible Mechanisms for Increasing Cardiac Intracellular Calcium Ion Concentration on Activation

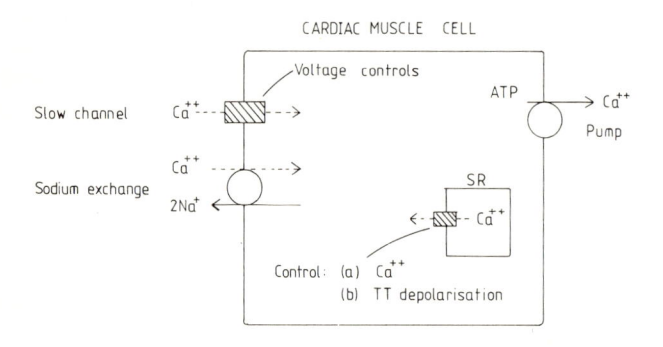

9.6 The Control of the Heart

The major control of the heart is by two systems of nerves which synapse on cardiac muscle carrying information from the central nervous system. These are the parasympathetic and sympathetic nerves. Parasympathetic nerves from the vagus synapse on the atria, including SA node, and release acetylcholine at their terminals. They cause a slowing of the heart rate or negative chronotropic effect (NCE). Sympathetic nerves synapse on all regions of the heart and release noradrenaline. They cause an increase in heart rate (positive chronotropic effect, PCE) and an increase in the force of cardiac contraction (positive inotropic effect, PIE). Using these nerves the heart can be made to respond to an increased workload. In exercise, for example, cardiac output in man can increase from 5 litres per minute to 30 litres per minute.

Acetylcholine causes an increase in the potassium conductance of the cardiac cell membrane. This results in a decrease in the rate of rise of the pacemaker potential, which tends to remain closer to V_K. Therefore the pacemaker potential takes longer to initiate the next action potential. This simple action of the transmitter appears to account satisfactorily for the negative chronotropic effect of the vagal parasympathetic nerves. The drug atropine is a competitive inhibitor of acetylcholine at the cardiac muscle membrane; it therefore increases heart rate when parasympathetic activity is present. A secondary and minor action of the parasympathetic nerves is to decrease the force of atrial contraction. This is probably also associated with the increased membrane potassium permeability, which encourages repolarisation and shortens the action potential, so that calcium ion entry is reduced.

Noradrenaline has approximately the same action on the heart as the related circulating hormone adrenaline. Adrenaline increases the inward calcium current during the plateau as shown in Figure 9.3 and thereby increases the tension generated (PIE). The origin of the positive chronotropic effect produced by adrenaline is not fully understood. In Purkinje fibres adrenaline produces a PCE by reducing potassium ion conductance during the pacemaker potential, producing a steeper rise. This mechanism does not operate at the SA node, where the PCE may be due to increased calcium ion entry during the pacemaker potential. As heart rate is increased it is necessary to reduce action potential duration so that the heart still has time to fill in diastole. Adrenaline accomplishes this by increasing the repolarising potassium ion current. Adrenaline and noradrenaline act on cardiac muscle by binding to membrane receptors known as β-adrenoreceptors. A group of drugs known as β-blockers are competitive inhibitors of the transmitter at these receptors; the best known of these drugs is propranolol. Propranolol inhibits the stimulating action of sympathetic nerves on the heart, and is used to prevent excessive demands being made on a diseased heart, and to reduce blood pressure. Unfortunately β-adrenoreceptors are found elsewhere in the body and considerable recent research has been devoted to the discovery of safe drugs selective for cardiac β-adrenoreceptors. The actions of acetylcholine and adrenaline on the cardiac action potential are summarised in Figure 9.6.

9.7 The Electrocardiogram

The bulk of cardiac muscle is excited in a particular spatial and temporal sequence due to the specialised conducting pathways of the heart. The extracellular currents produced by this excitation of an orientated muscle mass can be detected as voltage changes between different points on the body surface. Recordings are made from multiple electrodes placed on the skin of the limbs and chest wall, and the resulting traces are known as an electrocardiogram (or electrocardiograph) (ECG). A normal recording is shown in Figure 9.5. Five characteristic waves per heartbeat are seen, and these are designated P,Q,R,S and T. The relationship between the ECG and cardiac membrane currents is suggested by the atrial and ventricular action potentials drawn below the ECG. The P wave is associated with atrial depolarisation. The QRS complex is due to ventricular depolarisation and the T wave to ventricular repolarisation. Atrial repolarisation is lost in the QRS complex. The precise relationship between ionic currents and the ECG is complex and not fully understood, but the ECG is of great empirical importance in simple clinical diagnosis. First, although the currents due to the conducting system of the heart are too small to be detected, the temporal relationships between the ECG waves indicate how the conducting system is functioning. For example the time elapsed between the P wave and the R wave indicates the total conduction time through the AV node, down the bundle of His and along the Purkinje fibres into the ventricular muscle. The appearance of QRS complexes unrelated to P waves represents a loss of control by the conducting system over ventricular contraction. Various other types of abnormalities of the conducting system can be detected on the ECG. Secondly, a loss of function in one region of the heart is associated with a loss of the normal electrical activity in that region. Such loss of function is commonly due to an insufficient blood supply, as in a heart attack. The change in electrical activity produces characteristic alterations in the ECG. Furthermore different electrode positions record preferentially from certain areas of the heart, and this enables a localisation of the area of abnormality, as the ECG changes are more marked in some electrode configurations than others.

9.8 Summary

The excitable characteristics of cardiac muscle cells have been described and related to their pumping function. The cardiac muscle cell branches and makes contact with many nearby cells, with which it is electrically coupled. The cell shows spontaneous rhythmic electrical activity and can excite activity in its neighbours. Cells at the SA node have the highest rate of activity and act as the cardiac pacemaker. They excite the whole heart in sequence via a specialised conducting system. The cardiac action potential lasts several hundred milliseconds and its form is summarised in Figure 9.6. The initial spike is due to sodium ion entry; it is followed by a long plateau of calcium ion entry and a slow repolarising phase due to potassium ion efflux. The pacemaker potential results from a declining potassium ion conductance. The long refractory period of the cardiac action potential ensures that heart muscle will never generate a sustained tension. The electrical activity of cardiac muscle involves a specialised group of membrane ion channels of at least four types, which are gated by membrane voltage.

Figure 9.5: The Electrocardiograph and its Relationship to Atrial and Ventricular Action Potentials

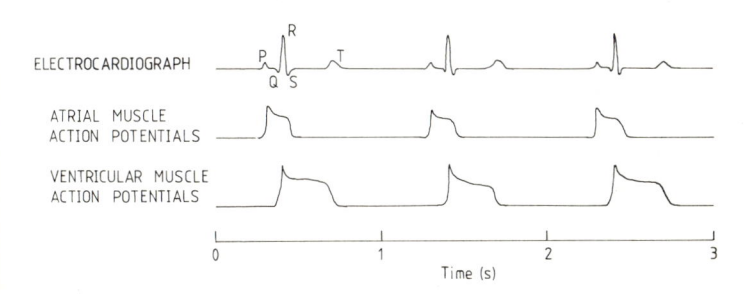

The mechanical activation of cardiac muscle is initiated by the calcium ion entry of the action potential plateau. This may stimulate further release of calcium from internal stores. Control of cardiac muscle is achieved by the neurotransmitters acetylcholine and noradrenaline. They alter the ionic currents across the cardiac muscle cell membrane. Acetylcholine slows the heart rate while noradrenaline increases both the rate and the force of contraction. The most important action of noradrenaline is to increase the calcium ion current into the muscle cell. Because cardiac function is closely related to electrical activity, the electrocardiogram provides an important, non-invasive aid for diagnosing heart disease.

For the purpose of comparison with skeletal and smooth muscle, the following features of cardiac muscle are of interest.

(1) Cardiac muscle exhibits rhythmic, electrical and mechanical activity in the absence of nerve stimulation. The cardiac nerves serve to alter the rate and force of activity.
(2) The cardiac action potential and contractile response are of similar time course, and a maintained tension is not possible.
(3) The transverse tubular system is poorly developed.

124 *Cardiac Muscle*

(4) Calcium ions for initiation of contraction come both from outside the cell, and from internal stores.
(5) There is direct electrical coupling between cells.

Figure 9.6: Summary of the Cardiac Action Potential

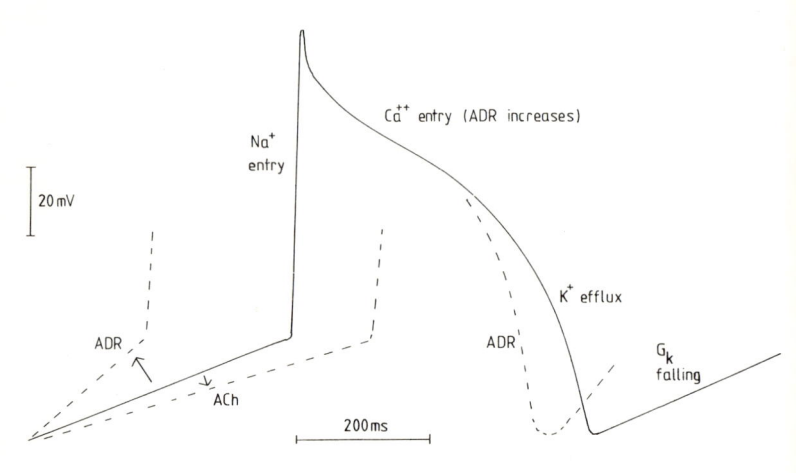

Key: ADR, adrenaline; ACh, acetylcholine.

Further Reading

Allen, D.G. and Blinks, J.R. (1978), Calcium transients in aequorin-injected frog cardiac muscle. *Nature 273,* 509.
Fabiato, A. and Fabiato, F. (1979), Calcium and cardiac excitation-contraction coupling. *Ann. Rev. Physiol. 41,* 473.
Fozzard, H.A. (1977), Heart: excitation-contraction coupling. *Ann. Rev. Physiol. 39,* 201.
McAllister, R.E., Noble, D. and Tsien, R.W. (1975), Reconstruction of the electrical activity of cardiac Purkinje fibres. *J. Physiol. 251,* 1.
Noble, D. (1979), *The Initiation of the Heart-beat.* 2nd edn, Clarendon Press, Oxford.
Reuter, H. (1979), Properties of two inward membrane currents in the heart. *Ann. Rev. Physiol. 41,* 413.
Schamroth, L. (1976), *An Introduction to Electrocardiography.* 5th edn, Blackwell, Oxford.

Vassalle, M. (1979), Electrogenesis of the plateau and pacemaker potential. *Ann. Rev. Physiol. 41,* 425.

Weidmann, S. (1974), Heart: electrophysiology. *Ann. Rev. Physiol. 36,* 155.

Weingart, R., Kass, R.S. and Tsien, R.W. (1978), Is digitalis inotropy associated with enhanced slow inward calcium current? *Nature 273,* 389.

10 SMOOTH MUSCLE

10.1 Distribution and Function

Smooth muscle is a third motor system, distinct from the cardiac and skeletal muscle systems. It differs from them in distribution, structure, mechanical properties and function. Most smooth muscle is found in the walls of hollow organs and tubular structures within the body. It is present, for example, in the walls of blood vessels, the alimentary tract, the respiratory tract and the urinary tract. Smooth muscle forms the bulk of the wall of the uterus and provides the contractile force for parturition. The functions of smooth muscle are manifold, but fall into two major groups. First, the diameter of tubular conduits may be regulated by the tension of the smooth muscle surrounding them. Secondly, a wave of smooth muscular contraction passing along a hollow organ can move the contents of the organ. This peristaltic pumping is used, for example, to propel food down the alimentary tract.

The control of smooth muscle is as vital to bodily function as that of cardiac or skeletal muscle. Consider the example of a man in heavy physical exercise. The violent contractions of his skeletal muscles are evident to the casual observer, while the increases in rate and force of the heart-beat are phenomena of common perception. Less obviously, the smooth muscles of his cardiovascular and respiratory systems are playing essential roles. The flow of blood to the active skeletal muscles is increased by relaxation of the smooth muscle walls around their blood vessels. Blood is temporarily diverted from non-essential organs, such as the skin and gut, by constriction of their blood vessels. The extra demand for oxygen is partly met by increasing the diameter of the airways of the lungs. Without these smooth muscle responses, sustained exertion would not be possible.

One difficulty in discussing smooth muscle is the variety of different behaviours exhibited by smooth muscles in different organs and species. The contractile response of smooth muscle is much slower than that of cardiac or skeletal muscle. It can exert a maintained tension for long periods, as is necessary, for example, to contain blood at high pressure within an artery. Functionally, two types of smooth muscle exist; single unit (or visceral) and multi-unit. The cells of multi-unit smooth muscle are each excited independently by a single nerve fibre and contract independently. The cells of visceral smooth muscle form a functional syncitium and contract as one unit; they do not each receive separate innervation. In this chapter the discussion will be confined to the visceral type which is the commoner of the two. Even amongst this type there are considerable variations in response and control. Many of the properties described below have been observed using smooth muscles from the large bowel known as the *taeniae coli*.

10.2 Structure and Contractile Response

Figure 10.1: A — Smooth Muscle Cells. B — Large Effect on Vessel Resistance due to Four-fold Change in Radius.

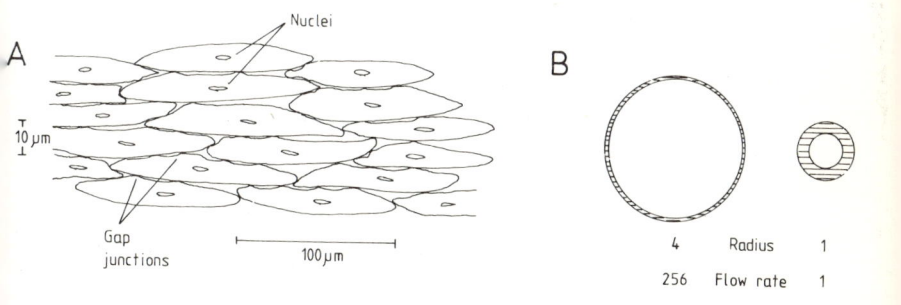

Smooth muscle cells are small and spindle-shaped, as drawn in Figure 10.1A. They are typically 5-10 μm in diameter at their widest point, and about 100 μm long. There is no transverse tubular structure and only a poorly-developed sarcoplasmic reticulum. Only inconspicuous striations are seen, but actin- and myosin-like proteins are present, and aligned approximately parallel to the fibre length. The cells fit closely together to form a network, and at many regions the membranes of adjacent cells approach very closely. These areas are known as nexuses or gap junctions. They are believed to be low resistance pathways between neighbouring cells. A broad correlation exists between the density of these junctions and the degree of electrical coupling between the cells. The nerve supply to visceral smooth muscle is variable in structure. In some tissues a proportion of the cells receive direct synaptic connections either from passing axons or from nerve terminals. In other tissues the nerves terminate at some distance from the muscle cells, and the transmitter substance released by each nerve must diffuse across a gap of several microns to influence a number of muscle cells. This long diffusion path is one factor associated with the slow contractile response of the muscle.

Contraction of smooth muscle involves a shortening of each cell without slippage between cells. The cell is able to exert tension over a large range of different lengths. It can contract between twice its resting length and half its resting length, giving a four-fold change overall. This is of considerable value in the control of flow rate in hollow organs. The conductance of a cylindrical tube to laminar fluid flow is proportional to the fourth power of its radius. Therefore a four-fold change in vessel circumference, mediated by mural smooth muscle, will alter the flow rate by a factor of 256, as shown in Figure 10.1B. The large range of smooth muscle contractibility consequently allows a powerful selective control of the blood flow to each tissue.

Visceral smooth muscle contracts in the absence of nervous or hormonal stimulation. Some tissues show a myogenic rhythmicity in which periods of contraction and relaxation alternate. These have a long time course (compared to cardiac muscle) of many seconds or minutes.

One unique feature of smooth muscle is that it can respond to stretching by increasing its tension, a mechanism by which it tends to maintain a constant length. The nerves supplying smooth muscle influence the tension and length of the fibre as well as the time period of oscillatory contractions. Nervous stimulation of a region of smooth muscle in the gut will increase the rate of rhythmic contraction, and increase the force of contraction. These slow contractile phenomena are controlled by the ion flows across the smooth muscle cell membrane.

10.3 Electrical Activity in Smooth Muscle

The intracellular free ionic concentrations in smooth muscle are difficult to measure. Internal sodium and potassium concentrations in taeniae coli are estimated to be 30 and 140 mM respectively, indicating their Nernst equilibrium potentials to be about 30 mV and −90 mV. The resting potential is approximately −55 mV. Use of the voltage-clamp technique to investigate ion flows in these cells presents several problems. The cells are small and of complex geometry. The extracellular spaces represent considerable resistances to current flow, and there is electrical coupling allowing direct current flow from cell to cell.

Most information has come from simple intracellular recording of the changes in membrane voltage associated with smooth muscle activity. Several types of electrical activity are seen, and some of these are illustrated in Figure 10.2. The spontaneous maintained tension is associated with myogenic trains of action potentials, as shown in traces A and B. Some cells show pacemaker potentials initiating these action potentials (A) and these cells are thought to drive the whole muscle mass through the cell to cell electrical coupling. The driven cells (B) fire at the same rate but have no pacemaker activity. Another form of electrical activity is the slow wave pattern shown in trace C. At the apex of each wave the membrane depolarisation is sufficient to evoke one or more action potentials.

Figure 10.2: Electrical Activity in Smooth Muscle Cells

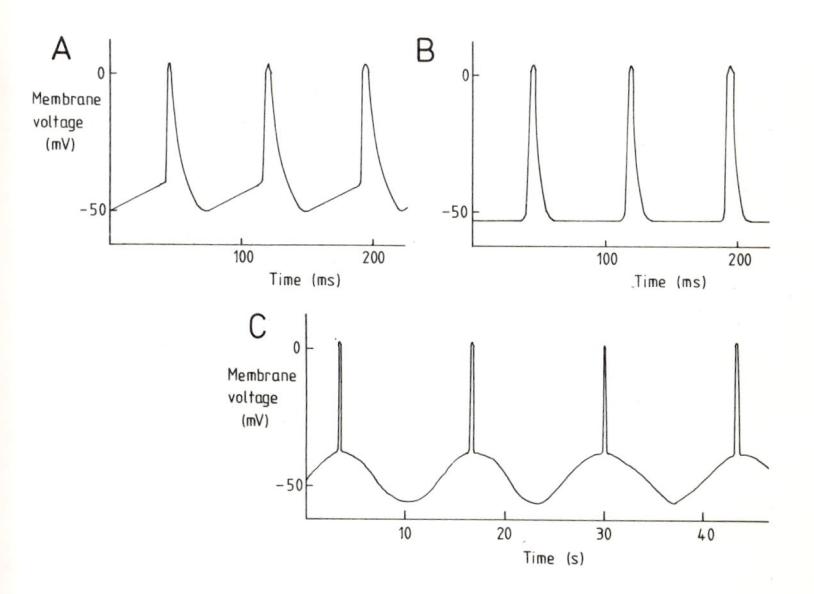

In guinea-pig taeniae coli the action potentials, or spikes, are insensitive to TTX and are dependent on external calcium ions. They disappear if external calcium is removed, and if calcium concentration is raised then the height and rate of rise of the spikes are both increased. The calcium-channel blocking drugs verapamil and D600 inhibit spike generation as do manganese ions. The spikes are due to a voltage-sensitive calcium channel in the smooth muscle membrane, which is opened by depolarisation and allows an influx of calcium ions to occur. Sodium ions have a dual influence on the action potential. On the one hand some of the inward current is carried by sodium ions, but on the other hand sodium ions also inhibit calcium entry. This inhibition results from a competition between sodium and calcium for some sites on the ion channels. Measures which inhibit spike generation also inhibit contraction, and the influx of calcium ions during each spike is important in the initiation of contraction.

The slow wave activity in certain smooth muscles generates rhythmic oscillatory contraction due to the superimposed spikes. The slow waves are not (except in the guinea pig) associated with alterations in the passive conductance of the cell membrane to ions. They are due to rhythmic variations in the activity of the membrane sodium pump. In smooth muscle this pump transports three sodium ions out of the cell for every two potassium ions pumped in. It is therefore electrogenic and drives a current out of the cell. Inhibition of the pump with ouabain is known to affect the resting potential, and it inhibits slow wave activity. Other methods of stopping the pump, such as addition of metabolic inhibitors like dinitrophenol or cyanide, also remove the slow waves. External potassium ions are essential for slow wave activity and the potassium ion concentrations required to stimulate pumping match those needed to produce slow waves. The mechanism by which the pump is induced to oscillate in activity is unknown.

Depolarisation of smooth muscle occurs on stretching, and this is the mechanism by which the cell resists lengthening. The depolarisation leads to an increased rate of action potential generation and consequently an increase in tension. Here the smooth muscle membrane is behaving like a mechanoreceptor. It appears to respond to mechanical deformation by increasing its conductance to sodium or calcium ions or to both. Neurotransmitters which excite smooth muscle also cause depolarisation, usually via a change in membrane ionic conductance.

10.4 Excitation-contraction Coupling in Smooth Muscle

In a number of smooth muscle preparations there is evidence that calcium ion entry associated with the action potential causes the initiation of contraction. In rabbit taeniae coli Seigman and Gordon (1972) have demonstrated that the relationship between external calcium ion concentration and twitch tension is of the form shown in Figure 10.3; external calcium ion concentration must exceed 0.5 mM for any contractile response. There is a sigmoid relationship between the tension and the logarithm of calcium concentration, with a saturation of tension response at about 6mM calcium. Exactly what proportion of the presumed rise in internal calcium ion concentration is due to influx through the action potential channel is not clear. There are certainly other mechanisms present which can increase free calcium levels in the cytoplasm. Chemical stimulation of smooth muscle contraction is still possible when the membrane is already depolarised. This indicates that there may be a calcium-entry channel opened by binding of chemical stimulants. The sarcoplasmic reticulum of smooth muscle stores calcium ions which may be released on stimulation. In the absence of a transverse tubular system this may be a form of calcium-induced calcium release. There is also evidence for a calcium/sodium exchange protein in the smooth muscle membrane which might be another mechanism of calcium ion entry. Figure 10.4 summarises some of these possible mechanisms for alteration of the internal free calcium ion concentration.

Figure 10.3: Rabbit Smooth Muscle Contraction Requires Extracellular Calcium Ions

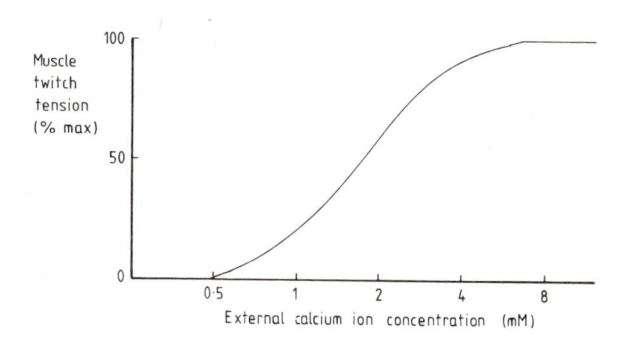

At present there is no unifying hypothesis of the mechanism of excitation-contraction coupling in smooth muscle. Excitation certainly leads to an increase in membrane calcium ion conductance and an influx of calcium, which is essential for contraction. Calcium release from internal stores may also be stimulated. Smooth muscle cells do not require a transverse tubular system as they are small enough for rapid diffusion to occur from the surface membrane to all internal structures. Relaxation of the muscle involves calcium binding to the cell membrane and to internal structures. In the long term there must be active calcium pumping both out of the cell and into the sarcoplasmic reticulum. Thus metabolic inhibitors cause a progressive rise in smooth muscle calcium content and fragments of smooth muscle membrane from guinea-pig ileum show ATP-dependent calcium ion binding.

Figure 10.4 Possible Mechanisms for Increasing Smooth Muscle Calcium Ion Concentration

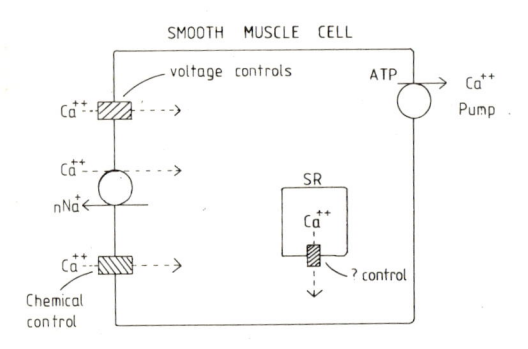

10.5 Nervous and Hormonal Control of Smooth Muscle

Smooth muscles of the visceral type can exert rhythmic or sustained tension in the absence of nervous or hormonal stimulation. These stimuli influence the magnitude of the tension and the rate of its rhythmic variation. The nerves which terminate in smooth muscle are part of the *autonomic* nervous system and are not usually under voluntary control. As in cardiac muscle, these motor nerves are divided into sympathetic and parasympathetic systems. Sympathetic nerves release noradrenaline as their chemical transmitter, while the parasympathetic transmitter is acetylcholine. Many smooth muscles receive both sympathetic and parasympathetic innervation and often the two inputs are antagonistic. The parasympathetic nerves usually stimulate contraction. For example gut movement is stimulated by the parasympathetic system and inhibited by sympathetic nerves.

Neurotransmitters bind to receptors on the smooth muscle membrane and thereby affect cell activity. The receptors are classified according to the transmitter to which they are sensitive. Thus acetylcholine binds to *cholinergic* receptors which are blocked by the drug atropine. Noradrenaline binds to *adrenergic* receptors which can be further subdivided into α- and β-receptors. Alpha-receptors are blocked by the drug phentolamine, and β-receptors are blocked by propranolol. The cardiac β-receptors mentioned in Chapter 9 are termed $β_1$ while the smooth muscle receptors are terms $β_2$. The mechanisms of transmitter action following receptor occupation vary from tissue to tissue. At different sites, the same transmitter may cause either contraction or relaxation. For example noradrenaline causes blood vessel constriction by an α-receptor action, which is inhibited by phentolamine. Noradrenaline also causes a relaxation of bronchiolar smooth muscle via a β action, which is blocked by propanolol. Even when the same receptor is employed, transmitter action may be quite dissimilar in two tissues. For example noradrenaline causes *relaxation* in taeniae coli via α-receptors.

At the molecular level chemical effects on smooth muscle often involve changes in membrane ionic conductance. Acetylcholine causes contraction in a number of tissues by increasing membrane conductance to sodium and calcium ions. The α actions of noradrenaline also involve a direct effect on membrane conductance. In taeniae coli noradrenaline increases potassium ion conductance, causing hyperpolarisation and relaxation. The β actions of noradrenaline are slower in onset and are associated with changes in the intracellular concentration of cyclic adenosine monophosphate (cAMP). The β-receptor appears to control the rate of cAMP production, and cAMP acts as a second messenger inside the cell.

A number of hormones influence smooth muscle, either acutely or over a prolonged period. Circulating adrenaline has actions similar to those of noradrenaline and acts over a period of minutes. The sex hormones oestrogen and progesterone exert a profound long-term effect on the uterine smooth muscle. In cats oestrogen acts to prime the uterus so that noradrenaline will excite it by an α action, while progesterone inhibits this and allows noradrenaline to inhibit contraction by a β action. Substances released locally, such as histamine and prostaglandins, are also important in the control of smooth muscle. This great variety of substances selectively influencing smooth muscle is of clinical importance, because the possibility exists of selective drug action by mimicking or blocking these substances. For example the drug salbutamol is a $β_2$-agonist; it has a relaxant noradrengeric action on the β-receptors of the respiratory tract, and causes bronchiolar dilatation. It is therefore useful in the treatment of some cases of asthma, where an inappropriate constriction of the airways occurs. Conversely the β-blocker propranolol, which is useful as a heart drug, inhibits noradrenaline in the lungs, causing airway constriction. Propranolol is therefore a dangerous drug to use on asthmatic patients.

10.6 Summary

Smooth muscle cells are characterised by their spindle-shape, lack of conspicuous striations and complete absence of a transverse tubular system. They are found in the walls of hollow internal organs and they control tubular diameter. Their contractions are slowly developing and tension can be maintained for long periods. Slow oscillations in tension occur, particularly in the smooth muscles of the gut. Spontaneous contractile activity is present associated with the spontaneous production of trains of action potentials. These spikes are primarily due to calcium ion entry through a voltage-sensitive channel. This calcium influx is one step in the initiation of contraction and further calcium ions may be released from stores in the sarcoplasmic reticulum. The slow rhythms of contraction are associated with slow electrical changes on which spike activity is superimposed. An electrogenic sodium pump is responsible for these electrical slow waves.

The rate and force of smooth muscle contraction is controlled by the sympathetic and parasympathetic divisions of the autonomic nervous system. These nerves release the transmitters noradrenaline and acetylcholine which bind to receptors on the muscle membrane. Some transmitter effects occur via alterations in membrane ionic conductances.

Different smooth muscles show wide variations in their properties, but the following points are of general interest in comparing visceral smooth muscle with skeletal and cardiac muscles.

(1) Smooth muscle, like cardiac muscle, shows spontaneous electrical and mechanical activity. Its nerve supply influences the rate and strength of contractions.
(2) The contractile response has an extremely long time-course and maintained tension is produced by trains of action potentials.
(3) There is no transverse tubular system.
(4) Calcium ion entry associated with the action potential spike initiates contraction.
(5) There is direct electrical coupling between cells.

Further Reading

Bolton, T.B. (1973), The permeability change produced by acetycholine in smooth muscle. In: Rang, HP. (ed.), *Drug Receptors.* University Park Press, Baltimore, 87.

Bulbring, E., Brading, A., Jones, A.W. and Tomita, T. (eds.) (1971), *Smooth Muscle.* Edward Arnold, London.

Prosser, C.L. (1974), Smooth muscle. *Ann. Rev. Physiol. 36,* 503.

Siegman, M.J. and Gordon, A.R. (1972), Potentiation of contraction: effects of calcium and caffeine on active state. *Am. J. Physiol. 222,* 1587.

Van Breemen, C., Aaronson, P. and Loutzenhiser, R. (1979), Sodium-calcium interactions in mammalian smooth muscle. *Pharmac. Rev. 30,* 167.

Part Five

THE CLINICAL IMPORTANCE OF EXCITABILITY

11 ANAESTHESIA

11.1 Introduction

In the preceding chapters the normal functioning of some excitable tissues has been described. We now turn to some examples of altered excitability which are of clinical relevance. Anaesthetics produce a reversible inhibition of excitable cell function and are divided by their clinical use into two groups. Local anaesthetics are injected close to peripheral nerves to cause a blockage of impulse transmission. General anaesthetics act on the central nervous system to produce a loss of consciousness and an inhibition of pain reflexes, allowing widespread surgical procedures to be performed. This chapter will describe the chemical and biological properties of some anaesthetics, and examine the extent to which their mechanisms of action are understood at the level of the excitable cell membrane. Significant strides have recently been made in research on these problems and this is partially a reflection of advances in knowledge of normal excitable cell function. Another important factor has been the development of various artificial models of the lipid bilayer portion of the cell membrane, which may be used to mimic the site of anaesthetic action.

11.2 Local Anaesthesia

A local anaesthetic (LA) is a substance which produces a reversible and dose-dependent inhibition of impulse conduction in a peripheral nerve, at low concentrations which are not irritant to the surrounding tissues. The chemical structures of two commonly-used local anaesthetics, lignocaine and procaine, are illustrated in Figure 11.1. Note that each molecule contains a hydrophobic section and a tertiary amine group which is hydrophilic; this amphipathic structure is a common feature of all local anaesthetics. The tertiary amines have pK values close to the pH of physiological fluids and so the cationic and neutral forms of the local anaesthetic will both be found in solution. The equilibrium between these forms is shown in Figure 11.1.

When applied to a nerve axon a local anaesthetic reduces the height, rate of rise and conduction velocity of the action potential. The threshold stimulus is increased but there is little or no effect on the resting potential. At higher concentrations of anaesthetic, impulse conduction is completely blocked, but the resting potential is maintained, and the block is reversed on removal of the anaesthetic. These effects are due to a blockage of the sodium and potassium channels of the nerve membrane, and a resultant suppression of the voltage-dependent ion flows which underlie the action potential. Voltage-clamp studies indicate that the sodium channel is more susceptible than the potassium channel to this inhibition. In the presence of the local anaesthetic a proportion of the sodium channels cease to function, and this proportion increases as the concentration of anaesthetic is raised.

Figure 11.1: The Structures of Two Local Anaesthetics

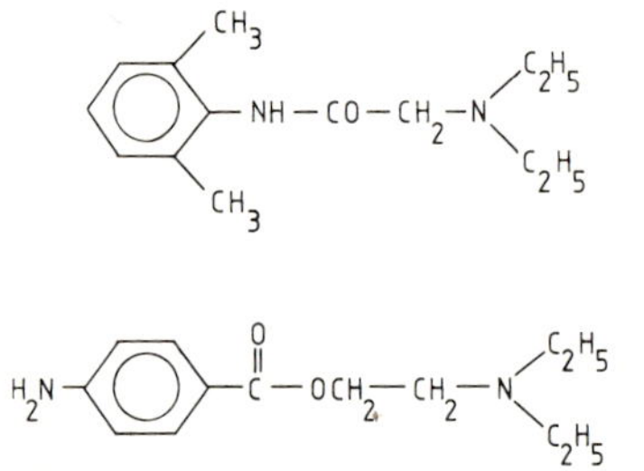

$$R_3N + H^+ \rightleftharpoons R_3NH^+$$

This channel blockade is most simply explained by the hypothesis that the local anaesthetic binds in the mouth of the channel and so prevents the passage of ions. In this model the LA is behaving like an ion which is too large to enter the pore. It is therefore of interest to know if it is the cationic or neutral form of the local anaesthetic molecule which is the active moiety. Experiments by Narahashi and colleagues (1970) and by Hille (1977) have shed considerable light on this question. It is possible to apply a local anaesthetic such as lignocaine in either its cationic or neutral form by varying the pH of the solution in which it is dissolved. At a pH of 6.0 the cation predominates while at a pH of 8.3 the neutral form is in excess. Alternatively the quaternary amino derivatives of these molecules are permanently cationic and should mimic the tertiary amine if the charged form is active. When applied to the outside of a single frog nerve fibre the neutral form of the anaesthetic is found to be more active and more rapid in action than the charged form. This result allows two interpretations. Either the local anaesthetic acts in the neutral form, or it is able to gain access to its site of action more readily when in that form. Further experiments suggest that the latter explanation is correct. For example when applied through the inside of the axon the charged form is always found to be the more active of the two. The anaesthetic must, it appears, cross a hydrophobic barrier to reach its site of action from the extracellular fluid, and this barrier is consistent in properties with the cell membrane. If the site of action is within the mouth of an ion channel then it is at the inner rather than the outer mouth of the pore.

Another observation is that the rate at which sodium channels are blocked by cationic local anaesthetic from within the axon is a function of the number of channels which are in the open state. The anaesthetic is most effective when the voltage-sensitive channel gates are open. This phenomenon, known as use-dependent inhibition, indicates that the anaesthetic is prevented from reaching its site of action if the gates are closed. The site is therefore held to lie within the channel itself and is separated from the intracellular medium by the channel gating structures. The anaesthetic can attain its site via a hydrophilic route only from the inside of the cell and only when the gates are open. The same site is also approachable via a hydrophobic route from either side of the cell membrane and channel blockage via this route is not a function of channel use.

Local anaesthetics at low concentrations block small-diameter nerve fibres before large-diameter fibres are affected. This is of clinical value as small-diameter pain fibres can be inhibited leaving larger motor fibres still functioning. At childbirth, epidural local anaesthetic can be used to block pain fibres but still leave the mother able to assist parturition by voluntary muscular activity. In other circumstances it is useful to inhibit both sensory and motor nerves and higher concentrations of local anaesthetic may be used to do this. Hernia repair may be accomplished under spinal local anaesthesia and here surgery is facilitated by paralysis of the muscles of the abdominal wall. The explanation for the high sensitivity of small-diameter axons to anaesthetics is related to the reasons for their low conduction velocity. Franz and Perry (1974) found that different 'critical lengths' of small and large nerves must be subjected to anaesthetic for a complete block of impulse conduction to occur. The smaller fibres had the shorter critical lengths.

As a small nerve has a larger resistance to the internal longitudinal current flow necessary for propagation, that current will fall off more rapidly with distance. If a short segment of axon contains no functioning ion channels then the internal current is less likely to 'leap' past the deficit in a small fibre. When local anaesthetic is applied near a nerve trunk containing large and small fibres then only a certain length of the trunk is subjected to anaesthetic concentrations high enough to block ion channels. If the length of that segment lies between the critical lengths for the large and small fibres then a differential block will occur. A higher concentration of applied anaesthetic will block all fibres as a longer segment of the nerve trunk will receive the required local concentration.

11.3 General Anaesthesia

Figure 11.2: Some General Anaesthetics

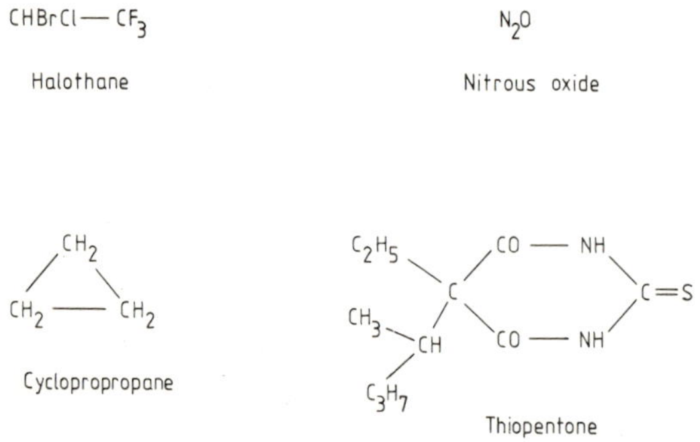

CHBrCl — CF$_3$

Halothane

N$_2$O

Nitrous oxide

Cyclopropropane

Thiopentone

General anaesthesia is a far more complex problem to unravel than local anaesthesia. One major difficulty is that there is no certain knowledge of either the anatomical or physiological site at which these substances act; it is consequently difficult to devise appropriate experiments to investigate the phenomenon at a sub-cellular level. A further complication is the enormous variety of chemical structures found in the group of molecules which cause general anaesthesia. Some of these molecules are illustrated in Figure 11.2 and, for example, there appears to be little in common between halothane and thiopentone, but they produce a similar clinical picture.

In the anaesthetised state there are changes in the patterns of neuronal activity in the CNS as measured in the electro-encephalogram (EEG). In particular there is a loss of cortical arousal and a depressed cortical response to sensory stimuli. A structure in the brainstem known as the reticular activating system (RAS) is important in stimulating the cortex and the present best guess is that general anaesthetics act on the RAS. The RAS contains many polysynaptic neuronal pathways which are thought to be particularly sensitive to interruption by general anaesthetics. At concentrations which cause clinical anaesthesia, general anaesthetics do not inhibit nerve impulse propagation but they can block synaptic transmission. The effects on the synapse vary from anaesthetic to anaesthetic, but both transmitter release and the postsynaptic response to transmitter are affected. As it is not clear which synapses in the brain are the sites for general anaesthesia, the relative importance of these two effects is not known. A further puzzle is that even during surgical anaesthesia some synaptic pathways in the CNS continue to function. Thus the regulation of blood pressure and respiration by the brain is maintained, although it is much more likely to require artifical assistance.

Historically an explanation for anaesthesia was first sought in the physico-chemical properties of anaesthetics in bulk solutions, and in correlations between these properties and the anaesthetic potencies of various molecules. One of the oldest correlations and the most successful is that of Meyer and Overton (see Meyer, 1899) who showed that the anaesthetic potency of a molecule is proportional to its olive oil-water partition coefficient. This relationship is valid over a potency range of ten thousand-fold and means that the more hydrophobic a molecule is, the more powerful an anaesthetic it makes. It follows from this that anaesthetics probably act from a hydrophobic site. Further deductions from the Meyer-Overton correlation are not possible as the brain is obviously not made of olive oil, and further progress demands the use of a more appropriate model for the site of anaesthetic action.

Given present knowledge of the mechanisms of nerve cell function, the anaesthetic site is almost certainly one or other of the components of an excitable cell membrane. Accordingly the hydrophobic region sought is either the interior of the lipid bilayer portion of the cell membrane, or it is a hydrophobic cleft in a membrane protein. The idea that membrane proteins are directly affected by general anaesthetics (as they are by local anaesthetics) is difficult to test experimentally. If nerve cell function is altered then the excitable ion channels are being influenced, but the anaesthetic action could be an indirect one via the lipid bilayer. Some membrane proteins, such as the sodium pump, are inhibited by general anaesthetics but only at far higher concentrations than those required to cause clinical anaesthesia. Other proteins such as haemoglobin and the light-emitting enzyme luciferase can be shown to bind general anaesthetics but no such binding to the excitable ion channels of the nerve membrane can be demonstrated. The hypothesis of direct protein binding is one which receives attention from time to time by exclusion of other hypotheses. Thus if evidence appears against a current theory of anaesthetic action via the lipid bilayer then the protein hypothesis regains popularity.

The theory of action from a site in the lipid bilayer has received closer experimental attention recently because techniques now exist for the construction of artificial lipid bilayers which are amenable to precise quantitative study. One model is the liposome which is a vesicle enclosed by one or many shells of bilayer; another model is the 'black film' which is a planar bilayer across which electrical measurements can be made. As expected anaesthetic absorption into the hydrophobic bilayer interior parallels biological potency. The Meyer-Overton correlation breaks down for large molecules, such as n-decane, which are still lipid soluble but nevertheless inert as anaesthetics. Haydon and co-workers have provided an explanation for this on the lipid bilayer model. They found that n-decane is too large to be significantly absorbed into a black film of cell membrane composition. The cut-off in the anaesthetic potency of the larger n-alkanes is paralleled by a cut-off in their partition into a lipid bilayer. This is significant evidence that anaesthetics of this type have a bilayer site of action.

If anaesthetics do act from the lipid bilayer then there must be some indirect mechanism by which the excitable membrane proteins are affected. The maintenance of a near-normal resting potential in the presence of general anaesthetics indicates that the bilayer structure has not been sufficiently disorganised to prevent it functioning as a barrier to ion flow. It must be failing in its second function of providing an appropriate environment for the membrane proteins. There are two hypotheses as to the mechanism of this failure: (1) that the fluidity of the bilayer interior is increased by anaesthetics and this inhibits protein function, or (2) that bilayer thickness and surface tension are increased by anaesthetics and this reduces the stability of membrane ion channels.

The degree of fluidity within a bilayer can be estimated by spectroscopic examination of certain probe molecules which enter the bilayer and report their own mobility. General anaesthetics cause a reversible increase in bilayer fluidity which is reversed by the application of hydrostatic pressure. The latter observation is important as general anaesthesia in mice, newts and tadpoles can be reversed by pressure. Changes in fluidity therefore appear to be well correlated with anaesthesia. However, changes in temperature also affect bilayer fluidity and a temperature rise of as little as $3°C$ will produce a fluidity change greater than that induced by anaesthetics at clinical concentrations. A simple interpretation of the fluidity hypothesis would therefore predict that a fever of $40°C$ should mimic general anaesthesia, which is clearly not the case. Furthermore the numbers of sodium and potassium ion channels functioning in the membrane of the squid axon is almost constant over a temperature range of $25°C$. Here large changes in fluidity do not appear to inhibit channel function. There may be some modified form of the fluidity hypothesis which could overcome these objections. It has been suggested that the fluidity of an annulus (ring) of lipid immediately surrounding the ion channel is altered by anaesthetics with deleterious consequences, but the evidence for this in nerve membranes is lacking.

Figure 11.3: The Thickness-tension Hypothesis. In the lower diagram the ion channel is unstable and likely to cease linking the aqueous phases.

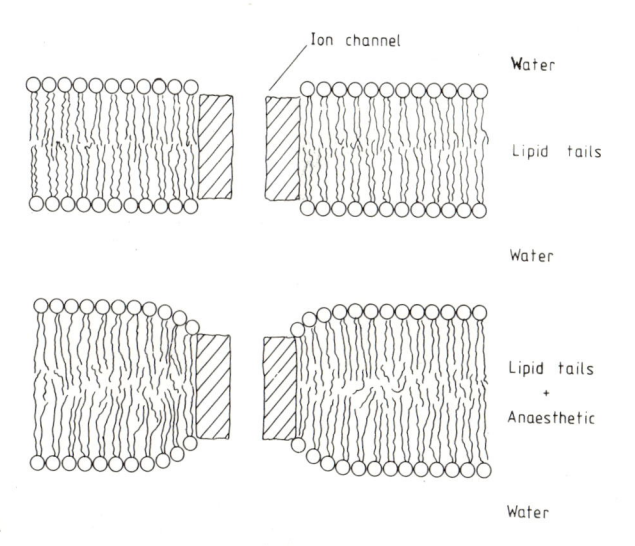

Small hydrocarbon anaesthetics such as n-pentane cause an increase in the thickness and tension of black films. Haydon has suggested that this might be relevant in anaesthesia. Excitable ion channels link the aqueous phases on either side of the membrane by spanning the hydrocarbon bilayer interior as shown in Figure 2.8. If the hydrocarbon interior is made thicker by an anaesthetic then the pore may find it increasingly difficult to link the aqueous phases, as an inward dimpling of the bilayer would be required as illustrated in Figure 11.3. Increases in bilayer tension make the dimpled structure more unfavourable energetically (i.e. of greater free energy). The pore-forming protein gramicidin A has been shown to be sensitive to bilayer thickness and tension in a manner which is quantitatively consistent with this dimpling model. For the thickness-tension hypothesis to be substantiated as a mechanism of anaesthesia it must be shown that anaesthetics cause a reversible thickening of nerve membranes and that the excitable ion channels are sensitive to such changes. At present the only method of measuring nerve membrane thickness is an indirect one involving the measurement of membrane electrical capacity. As explained in Chapter 2 (page 16) a decrease in membrane electrical capacity indicates an increase in membrane thickness, if certain assumptions are made. Pentane causes a reversible decrease in the high-frequency capacity of the squid axon membrane and so appears to cause thickening. This effect is accompanied by a reversible decrease in the number of functioning sodium channels. The thickness-tension mechanism is therefore well-supported as an explanation for the local anaesthetic action of pentane.

The relevance of this mechanism to general anaesthesia due to clinically-important molecules is less clear. First the sodium channel of the nerve membrane is unlikely to be the site of general anaesthetic action. The concentrations of general anaesthetic required to block the nerve are too high. Secondly pentane and, for example, halothane differ in their effects on the nerve and may act in different ways both on the nerve axon and in the brain.

General anaesthesia remains something of a mystery. It is of considerable theoretical interest to know how the excitable cell is depressed by such molecules. There is also a practical aspect to this interest as general anaesthesia is a dangerous procedure. It is a particular risk in the very young, in the old and in those with respiratory or cardiovascular disease. Some anaesthetics are much safer than others and the ideal anaesthetic has not yet been discovered. An understanding of the mechanisms of anaesthesia and of the mechanisms of anaesthetic toxicity would undoubtedly lead to safer anaesthetic procedures.

Further Reading

Fink, B.R. (ed.) (1975), *Molecular Mechanisms of Anaesthesia*. Raven Press, New York.
Franz, D.N. and Perry, R.S. (1974), Mechanisms for differential block among single myelinated and non-myelinated axons by procaine. *J. Physiol. 236,* 193.
Haydon, D.A. (1975), Functions of the lipid in bilayer ion permeability. *Ann. N.B. Acad. Sci. 264,* 2.
Haydon, D.A., Hendry, B.M., Levinson, S.R. and Requena, J. (1977), The molecular mechanisms of anaesthesia. *Nature 268,* 356.
Hendry, B.M., Urban, B.W. and Haydon, D.A. (1978), The blockage of the electrical conductance in a pore-containing membrane by the n-alkanes. *Biochim. Biophys. Acta 513,* 106.
Hille, B. (1977), Local anaesthetics: hydrophilic and hydrophobic pathways for the drug-receptor interaction. *J. Gen. Physiol. 69,* 497.
Lee, A.G. (1976), Model for action of local anaesthetics. *Nature 262,* 545.
Meyer, H. (1899), Zur theorie der alkoholnarkose. Welche eigenschaft der anasthetica bedingt ihre narkotische wirkung? *Arch. Exp. Pathol. Pharmakol. 42,* 109.
Miller, K.W., Paton, W.D.M., Smith, R.A. and Smith, E.B. (1973), The pressure reversal of general anaesthesia and the critical volume hypothesis. *Mol. Pharmacol. 9,* 131.
Mullins, L.J. (1971), Anaesthetics. In: Lajtha, A. (ed.), *Handbook of Neurochemistry,* Vol. VI. Plenum Press, New York, 395.
Narahashi, I., Frazier, D.T. and Yamada, M. (1970), The site of action and active form of local anaesthetics. *J. Pharm. Exp. Ther. 171,* 32.

12 MEMBRANE EXCITABILITY AND DISEASE

12.1 Introduction

Many diseases are associated with disorders of excitable cell function. In some cases there is no primary defect in the excitable cell, but it finds itself in an unsuitable environment due to a distant disease process. For example, persistent diarrhoea and vomiting due to infection may lead to a loss of body potassium, and to a low extracellular potassium concentration. The excitable mechanism is critically dependent on transmembrane potassium ion gradients; it will therefore malfunction as a secondary effect of the illness. Other diseases involve a primary defect in excitability. This group includes epilepsy, Parkinson's disease, multiple sclerosis, myasthenia gravis, myotonia, botulism and tetanus. Psychiatric diseases such as depression, schizophrenia and anxiety states are certainly associated with altered excitable cell function, but are poorly understood at the cellular level.

Primary diseases of excitable cells outside the central nervous system are the simplest to study. Thus it is that myasthenia gravis and myotonia are at least partially understood at the level of the cell membrane. Secondary excitable cell malfunction is liable to both central and peripheral manifestations; the latter are better understood. Diseases of the central nervous system are the most difficult of all to unravel. In this chapter most of the discussion will be confined to a few conditions where at least some sub-cellular insight is possible.

12.2 Myasthenia Gravis

Myasthenia gravis is a disease of muscle characterised by an abnormal fatiguability. It usually presents between the ages of 20 and 50 and is more common in females than males. The initial symptom is often a weakness of the ocular muscles, resulting in a ptosis or drooping of the eyelids. Other muscles become involved, producing difficulty in swallowing and speech impairment. The fatiguability is evident in the decline in articulation during a conversation. Symptoms also become worse towards the end of the day. On clinical examination at this stage a weakness of the facial muscles is common. The disease runs a variable chronic course with many relapses and remissions; there is often a progression of symptoms to affect a widespread group of muscles. Pregnancy and infection are reported to predispose to a myasthenic crisis in which there is an acute exacerbation of the weakness. In the later stages of the disease, involvement of the muscles of respiration is important and dyspnoea is often present. There is an increased risk of bronchopneumonia and of death due to cardiac or respiratory failure.

Skeletal muscle biopsy in myasthenic patients reveals abnormalities of the endplate and muscle cell. There is a decrease in the electrical response of the endplate to acetylcholine. This is temporarily relieved by the acetylcholinesterase inhibitor neostigmine. The spontaneous miniature endplate potentials are smaller than in control muscles, although the nerve terminal transmitter vesicles appear normal. Direct binding studies show a decreased number of acetylcholine receptor sites at the endplate. These findings are correlated with the presence of antibodies in the plasma which react with the acetylcholine receptor. The disease is therefore considered to be due to the inappropriate blocking of acetylcholine receptors by antibodies produced by the body's immune system.

The causes of this unfortunate autoimmune attack are not clear. Many cases of myasthenia are associated with an enlarged thymus, and the plasma also contains antibodies directed against thymic epithelial cells. One hypothesis is that lymphocytes become sensitised to thymic cells (by an unknown mechanism) and that the antibodies so produced cross-react with skeletal muscle membrane proteins. Myasthenia occasionally also occurs as a symptom in other conditions, such as carcinoma, collagen disease or thyroid disease. Alterations in the immune system are also the most likely cause of this 'secondary' myasthenia. The selective effect of the disease on certain muscle groups is interesting. It implies that there may be differences in the structure and immune specificity of the acetylcholine receptors of different skeletal muscles in the same individual.

The diagnosis of myasthenia rests on the demonstration of transient symptomatic relief by a short-acting anti-cholinesterase such as edrophonium (Tensilon). Electromyography also shows a characteristic improvement due to edrophonium. The treatment of myasthenia is with a long-acting inhibitor of cholinesterase (pyridostigmine). This does not always produce complete remission. There are side-effects due to excessive autonomic cholinergic activity (abdominal colic, salivation and cardiac arrythmias) which may require treatment with atropine. On the basis that the disease is autoimmune, and may be improved by immunosuppression, steroids and adrenocorticotrophic hormone (ACTH) have recently been tried therapeutically. They are promising but, as yet, unproven treatments. Thymectomy is also reported to cause remission.

12.3 Dystrophia Myotonica

This is a rare hereditary muscular dystrophy, transmitted as a dominant trait but with poor penetrance. Myotonia is the inability of muscles to relax properly after contraction. The disease itself is a complex clinical syndrome which, in addition to myotonia, includes muscle weakness and wasting, cataract and gonadal atrophy. Symptoms first appear between the ages of 15 and 40. Sudden falling is a common presenting complaint, due to muscular weakness. There is wasting of the facial muscles, sternomastoids, shoulder, arm and leg muscles. The prolonged after-contraction is most commonly found in the hands which are unable to relax their grasp. Myotonia is exacerbated by cold, fatigue and emotion.

A similar hereditary disorder exists in goats. This has allowed investigation of the electrophysiological basis of myotonia, and the results appear to be applicable to the human disease. Bryant (1969) compared the passive electrical properties of muscle membranes from myotonic and non-myotonic goats. The major difference is that the resting electrical conductance of the membrane is much reduced in myotonia, from 500 μS/cm^2 to about 180 μS/cm^2. This difference is accounted for by a loss of membrane conductance to chloride ions. Adrian and Bryant (1974) demonstrated that repetitive firing, as in myotonia, occured in normal muscle fibres in chloride-free Ringer solution. It appears that movement of chloride ions into the muscle cell is an essential component of the current flow by which the cell recovers from a depolarisation. This chloride influx cannot take place if external chloride ions are removed, or if there is a defect in the membrane chloride conductance. Each muscle action potential produces a movement of potassium ions out of the cell into the transverse tubules, where they are trapped. Such an accumulation of tubular potassium ions leads to a progressive rise in the resting potential; this rise is made greater by the low ionic conductance of the surface membrane. The resting potential exceeds the threshold for action potential generation, and repetitive firing occurs. This behaviour can be predicted with a mathematical model of the muscle cell membrane in which the chloride conductance of the surface membrane is one-tenth of its normal value.

The prognosis in dystrophia myotonica is extremely variable. It does not shorten life-expectancy and indeed there is sometimes a slight improvement in old age. Myotonia is associated with myocardial involvement and abnormalities in the conducting system of the heart have been reported. There is no treatment for the muscular weakness and wasting, which are usually the major features of the disease. The myotonia itself is seldom severe enough to require treatment but phenytoin can be used for this. Phenytoin is a drug more often employed as an anti-convulsant, and its mode of action in myotonia is unknown.

12.4 Plasma Electrolyte Disturbances

The concentrations of potassium and calcium ions in human plasma (or extracellular fluid) are relatively low, compared to the total electrolyte concentration. Plasma potassium usually lies in the range 3.5-5.0 mM. Plasma calcium has normal values between 2.2 and 2.6 mM. It follows that small changes in their absolute concentrations can markedly affect their chemical potential and have a profound influence on cellular function (due to the logarithmic relationship between concentration and chemical free energy). For excitable cell function the important variable is the transmembrane chemical gradient for each ion. Unfortunately it is not possible in routine clinical practice to measure intra-cellular ionic concentrations. Nevertheless there are recognised clinical consequences of altered extracellular potassium and calcium ion concentrations, which are consistent with experimental data in animals and which are amenable to treatment.

As potassium is primarily an intracellular ion, the plasma concentration is *not* a good guide to the total body potassium balance. Low plasma potassium concentration may follow excessive excretion of the ion. This occurs (1) in prolonged diarrhoea and vomiting, (2) following diuretic therapy with drugs which increase the renal excretion of potassium (thiazides), and (3) in certain endocrine diseases such as Cushing's disease and aldosteronism. The clinical picture is of weakness and lethargy. Electrocardiography shows characteristic changes, with flattened T waves and sometimes an extra wave associated with repolarisation known as the U wave. The treatment must be directed against the cause of the imbalance; in addition oral or intravenous supplements of potassium may be required and their effects on plasma potassium should be carefully monitored. High plasma potassium can occur (1) in renal failure, (2) due to excessive intake, (3) in acidosis and (4) in Addison's disease. Again weakness is a common clinical feature but the most dangerous complication is of electrical failure of cardiac muscle. The electrocardiogram demonstrates large T waves, and may progress to show the usually fatal condition of ventricular fibrillation. The control of the repolarising phase of the cardiac action potential is intimately dependent upon potassium ion currents. This phase is monitored in the ECG as the T wave, hence the characteristic effects of potassium on the ECG. If disruption of these currents is severe, then the control of cardiac rhythm is completely lost, and an ineffective fibrillation results. Plasma potassium ion concentrations in the range 6.0 to 8.0 mM require urgent treatment. Intravenous calcium gluconate and insulin in dextrose are effective in reducing the potassium level. Dialysis may be considered in some cases.

Calcium ions exert a controlling influence on cellular excitability. High extracellular calcium concentrations tend to reduce excitability; low concentrations lead to a state of 'hyperexcitability' in which the slightest stimulus will cause excitation. This effect is quite separate from the role of calcium ion currents in phenomena such as excitation-secretion coupling and excitation-contraction coupling. The calcium ion appears to have a profound influence on the threshold for excitation. This is due to binding to extracellular membrane sites and not due to a transmembrane calcium current. The interested reader is referred to the seminal work of Frankenhaeuser and Hodgkin (1957) and to the mathematical description of the effect provided by Huxley (1959).

In health the plasma calcium concentration is regulated by parathyroid hormone; an excess of this hormone, in the disease of hyperparathyroidism, leads to hypercalcaemia. High plasma calcium may be produced by primary or secondary neoplasms in bone; it is also found in association with diseases such as TB, sarcoidosis, thyrotoxicosis and Addison's disease. The increased threshold for cellular excitation is manifest clinically as weakness and depression. Other features of hypercalcaemia are thirst and polyuria, constipation, anorexia and vomiting. Very high calcium concentrations may lead to coma and death. Acute treatment is based on measures to increase calcium excretion and inhibit calcium reabsorption from bone. A saline diuresis is simple the *calcitonin* may also be required. Peritoneal dialysis is indicated in the aggressive management of severe cases. Dexamethasone is sometimes effective in suppressing calcium reabsorption. Definitive treatment is of the underlying disorder. In hyperparathyroidism this usually involves surgical removal of most of the parathyroid tissue.

Low plasma calcium ion concentration may result from hypoparathyroidism; it may also be a feature of chronic renal failure or occur due to a dietary defficiency (particularly in lactation). The clinical manifestations of the low threshold for cellular excitation include, (1) pins and needles in the lips, nose and fingers (paraesthesiae); (2) spontaneous muscle cramps; (3) epileptic fits; and (4) a positive Chovstek's sign, in which tapping the facial nerve beneath the zygoma leads to a twitch of half the face. This clinical syndrome is known as tetany. The acute symptoms and signs will respond to a slow intravenous injection of calcium gluconate. Further treatment may include a high calcium diet. Hypoparathyroidism is treated with vitamin D which increases the absorption of calcium from the gut.

12.5 Diseases of the Central Nervous System

The complexity of organisation in the CNS is such as to make investigation of disease within it a daunting prospect. This is particularly so in the case of those diseases whose clinical manifestations are sufficiently complex and antisocial for them to be labelled as psychiatric. We cannot yet speak of the biochemical basis of schizophrenia or depression. Even accepting these difficulties, there has been very little progress in research into many severe nervous diseases. Multiple sclerosis, for example, is a tragic disease in which the pathological lesion is easily demonstrable. There is a disseminated demyelination of neurones within the CNS, producing widespread sensory and motor disabilities. However, the aetiology of this process remains obscure and little rational therapy can be offered to these patients. Epilepsy, on the other hand, is now a treatable problem. It results from an abnormal focus of excitation in the CNS, which discharges to produce a pathological pattern of neuronal firing. The precise clinical manifestations of the discharge depend on its position in the cerebrum. A focus in the motor cortex will produce a characteristic inco-ordinated fit; one in the temporal lobe tends to initiate a co-ordinated, but inappropriate, pattern of activity Clinical neurologists are now armed with a powerful and effective group of anti-convulsant drugs which inhibit the initiation of the discharge. The development of these drugs has largely been a matter of trial-and-error, but they provide a very successful therapy. In animal models they appear to inhibit neuronal excitation by a presynaptic action.

Parkinson's disease involves the CNS lesion which is probably the best understood at the molecular level. It is a disturbance of voluntary motor function manifested as the clinical triad of rigidity, tremor and bradykinesia (slow movements). The walk is characteristic; the arms do not swing fully and the gait is shuffling. The face is expressionless and speech is dysarthric (poorly articulated) and monotonous. Handwriting is commonly affected early in the disease and becomes progressively smaller and less legible. The tremor affects the hands; it is worst at rest and in anxiety, but is improved by voluntary movement. Examination classically reveals a 'cogwheel' rigidity at many joints. In its late stages the disease is extremely disabling. It is, however, a reversible phenomenon. Sufferers have been reported to run out of burning houses. One even assisted in the rescue of victims following a road accident before relapsing into immobility.

The biochemical abnormality in Parkinson's disease is a low level of the neurotransmitter dopamine in the basal ganglia. The ganglia are centres in the telencephalon for the control of motor function. They receive neuronal inputs from many other centres. One input is from the substantia nigra of the midbrain, via inhibitory nerves releasing dopamine from their terminals. In Parkinsonism these terminals are severely depleted of dopamine and the inhibitory effect of the substantia nigra on the basal ganglia is lost. Excitatory inputs to the basal ganglia from other centres (releasing acetylcholine at their terminals) result in excessive activity and the characteristic disturbance of voluntary motor function. Two possible therapeutic methods are theoretically justified by this description. First the use of anti-cholinergic drugs to block the unrestrained excitation of the basal ganglia. Secondly stimulation of the dopaminergic pathways by l-dopa, which is a precursor of dopamine and stimulates dopamine synthesis. Both methods have proved useful but the second is more recent and more effective. The decarboxylase inhibitor carbidopa prevents the extracerebral breakdown of l-dopa and leaves more available for transport into the brain. These two drugs can be given together in the form of Sinemet, allowing a smaller total administration of l-dopa to be effective.

In about 20 per cent of patients with Parkinson's disease the response to l-dopa is dramatic. Most patients experience considerable symptomatic relief. The drug sometimes has the severe side effects of anorexia, nausea, vomiting, hypotension and dyskinesia (writhing of the mouth and tongue). Maintaining a balance between the therapeutic and toxic effects of 1-dopa by alteration of dosage can be an impossible task. The treatment is certainly not a cure. Prolonged drug administration in Parkinsonism may not prevent a deterioration of the condition to the point where drugs appear no longer to be of any use at all. In the worst cases, rapid transitions occur between the symptoms of Parkinsonism and 1-dopa overdosage. Clearly the treatment of a local biochemical abnormality with systemic drug therapy is unlikely to prove completely successful. The CNS employs the same transmitter systems in many different locations; even a drug whose action is specific to one transmitter system is likely to have a multiplicity of actions, some of them unwanted. In practice further difficulties exist. Drugs are not even specific for a single transmitter system but tend to cross-react with other synaptic processes. Finally the CNS appears to respond to the chronic administration of drugs by becoming progressively less sensitive to them.

Further Reading (Chapter 12 and General)

Adrian, R.H. and Bryant, S.H. (1974), On the repetitive discharge in myotonic muscle fibres. *J. Physiol. 240*, 505.

Adrian, R.H. and Marshall, M.W. (1976), Action potentials reconstructed in normal and myotonic muscle fibres. *J. Physiol. 258*, 125.

Bolis, L., Hoffman, J.F. and Leaf, A. (1976), *Membranes and Disease.* Raven Press, New York.

Bryant, S.H. (1969), Cable properties of external intercostal muscle fibres from myotonic and non-myotonic goats. *J. Physiol. 204*, 539.

Edelman, G.M. and Mountcastle, V.B. (1978), *The Mindful Brain.* MIT Press, Cambridge, Mass.

Frankenhaeuser, B. and Hodgkin, A.L. (1957), The action of calcium on the electrical properties of squid axons. *J. Physiol. 137*, 218.

Harper, P.S. (1979), *Myotonic Dystrophy.* W.B. Saunders Co., London.

Huxley, A.F. (1959), Ion movements during nerve activity. *Ann. N.Y. Acad. Sci. 81*, 221.

Kuffler, S.W. and Nichols, J.G. (1976), *From Neuron to Brain.* Sinauer Associates, Sunderland, Mass.

Mathews, W.B. and Miller, H. (1975), *Diseases of the Nervous System.* Blackwell Oxford.

Porter, R. (ed.), (1978), *Studies in Neurophysiology.* Cambridge University Press, Cambridge.

Weissmann, G. and Claiborne, R. (eds.) (1975), *Cell Membranes: Biochemistry, Cell Biology and Pathology.* HP Publishing Co., New York.

INDEX

absorption spectrum of photopigments 77
accommodation in mechanoreceptors 93
acetycholine (ACh)
   in cardiac muscle 120, 123
   in smooth muscle 132
   synaptic transmitter 59
acetylcholinesterase (AChE) 59
acetylcholine-sensitive channel 64
action potential 42
   propagation of 51
   sensory initiation of 92
action spectrum of visual pigments 76
active membrane in nerve 47
active transport 21, 31
adenosine triphosphate (ATP) 21
adrenaline
   and cardiac muscle 118, 121
   and mechanoreception 91
adrenoreceptors 121, 133
aequorin 104, 119
amacrine cells 74
atrio-ventricular (AV) node 113
atropine
   in cardiac muscle 121
   in smooth muscle 133
autonomic nervous system 132

bacteriorhodopsin 20, 78
bilayer
   capacitance of 16, 101, 146
   lipid 15
   permeability of 17
bipolar cells of retina 74
black film 144
   and bilayer capacitance 16, 146
botulism 148
bundle of His 113

calcium ion channels
   at synapse 59, 61, 68
   of cardiac muscle 117-18
   of smooth muscle 129
calcium ion pump 31, 61, 105-6
cardiac action potentials 115-17

cardiac pacemaker 112
carrier protein 22
central neurotransmitters 65
chemical energy 26
chemical transmission 57
   vesicular storage in 58
Chovstek's sign 153
colour blindness 84
colour vision 82
conducting system of the heart 113
conduction velocity 52
cones 73-4
   pigments of 83
crayfish stretch receptor 91
curare 65
cyclic adenosine monophosphate (cAMP) 133
cyclic guanosine monophosphate (cGMP 81

D600 118, 129
dark currents in rods 79
depolarisation in nerve 47
diastole 114
diuretic therapy 151
dopamine in Parkinsonism 154

early receptor potential (ERP) 78
edrophonium 149
EGTA and calcium 104
electrical capacitance
   and anaesthesia 146
   in skeletal muscle 101
   of bilayers 16
electrical energy 26
electrical synapses 67
electrocardiogram (ECG) 113, 121
electrochemical gradient 21, 26
electro-encephalogram (EEG) 143
electron density profile 15
endplate potential (epp) 63
epilepsy 153
equilibrium potential 29
excitation-contraction coupling 109, 131
excitation-secretion coupling 60
exocytosis 59
extrinsic membrane proteins 19

ferritin 101
fovea of retina 74
free energy 27
frog neuromuscular junction 57

ganglion cell of retina 74, 82
gated ion channels 35, *see* calcium
 ion channels, potassium
 channel, sodium channel
general anaesthesia 142
generator potential 89-90
giant axon of squid 41
gramicidin 23, 64, 146

Halobium halobius 19, 78
halothane 142, 146
horizontal cells of retina 74
hypercalcaemia 152
hyperparathyroidism 152
hypocalcaemia 153
hypoparathyroidism 153

image potential barrier 17
inhibitory postsynaptic potential
 (ipsp) 66
inhibitory synapses 66
intercalated discs
 of cardiac muscle 113
intrinsic membrane proteins 19
inward rectification
 of potassium conductance 104
ion pumps 21, 31, *see* calcium ion
 pump, sodium pump

local anaesthetics 53, 139
local circuit currents 51
L-DOPA 154
light current in rods 79
lignocaine 139
lipid bilayer 15-17, *see* membrane
liposome — artificial bilayer 144
lymphocyte capping 18

membrane
 capacitance 16, 101, 146
 composition 14
 conductance 34
 current noise 64
 fluidity 144-5
 proteins 18
Meyer-Overton correlation 143
miniature endplate potentials
 (mepps) 63, 149
motor unit 99
multiple sclerosis 153
myasthenia gravis 148
myelin sheath 15, 52
myotonia 104, 148, 150

negative chronotropic effect (NCE)
 120
neostigmine 149
Nernst equation 29
Nernst potential 46, 102, 116, 125
nerve
 axon 41
 impulse propagation 52
 squid 42
nodes of Ranvier 52
noradrenaline 121, 123

omega figure 59
optic nerve 73
ouabain 45, 54

pacemaker potential 115-16, 129
pacinian corpuscle 87-90
parathyroid hormone 152
Parkinson's disease 148, 154
pentane, anaesthetic
 properties of 146
peripheral receptive field 88
phentolamine 133
phosphatidylcholine 14
phosphodiesterase 81
photoisomerisation 80
photopic vision 84
photoreceptors 73-4
phototransduction 73
pore protein 22
positive chronotropic effect (PCE)
 120
potassium channel 49
 and anaesthetics 140
 in skeletal muscle 104
pressure reversal of anaesthesia 145
procaine 92, 107, 139
propranolol 65, 133-4
purkinje fibres 113-16
'purple' membrane 19
pyridostigmine 149

red blood cell membrane proteins 19
refractory period 43, 51, 117

Index

regenerative mechanism 37
resting permeability 22, 32
resting potential 33, 34, 46
reticular activating system (RAS) 143
retina 73
retinal 76
reversal potential 63, 91, 118
rhodopsin 75
rods 73-4

sacromere 100
saltatory conduction 53
sarcoplasmic reticulum (SR) 101, 105, 115
    coupling to TT 107
Schwann cells 52
scotopic vision 84
sensory adaptation 89, 92
Sinemet, in Parkinsonism 154
sino-artrial (SA) node 112
slow wave activity 129-30
sodium channel
    and anaesthetics 140
    in cardiac muscle 117
    in nerve 49, 55
sodium hypothesis 48
sodium pump 21, 44-6, 130
    and ouabain 45, 54
spatial summation 67
squid giant axons 42
staircase effect 119
succinylcholine 65
systole 114

taeniae coli 127-35
temporal summation 67
tetanus 148
tetany 153
tetraethylammonium (TEA) 49
tetrodotoxin (TTX) 49, 55, 90, 103, 117, 129
thickness-tension hypothesis of anaesthesia 146
thiopentone 142
threshold stimulus 37, 43, 50
thymus, and myasthenia 149
transmitter substance 57-8, *see* acetylcholine, noradrenaline
transverse tubules (TT) 100-3, 115
    coupling to SR 107
trichromatic theory 83
troponin 106

use-dependent inhibition by anaesthetics 141

valinomycin 23
verapamil 118, 129
visual sensitivity 81
vitamin A 76
voltage clamp 48, 117, 128
voluntary movement 99

X-ray diffraction 15

Z-lines 100, 103